CONGENITAL MALFORMATIONS

CONGENITAL MALFORMATIONS

Case Studies in Developmental Anatomy

MARTHA E. SUCHESTON, Ph.D.
Department of Anatomy
College of Medicine
Ohio State University

M. SAMUEL CANNON, Ph.D.
Department of Anatomy
School of Medicine
University of Texas, Galveston

Photography by
GABRIEL A. PALKUTI, R.B.P.
Department of Anatomy
College of Medicine
Ohio State University

F. A. DAVIS COMPANY, Philadelphia

ISBN 0-8036-8210-7

Library of Congress Catalog Card Number 73-77111

Printed in the United States of America.

To Louis
for love and encouragement

To my Family and Friends

OF ALL THE TRIBULATIONS OF PHYSICAL EXISTENCE, THE ONE MAN WILL SURELY NEVER HAVE TO UNDERGO AGAIN IS THE HAZARD OF BEING A FETUS. UNKNOWN

Preface

In preclinical anatomy the student must master myriad information at a time when he is unaware of its application. Developmental anatomy provides the logical basis for all anatomical disciplines, particularly gross anatomy, and an explanation, however superficial, of many congenital malformations. The purpose of this book is to present case studies of selected congenital defects, those which best demonstrate basic developmental concepts and which emphasize anatomical reasoning and the application of anatomical principles to clinical signs. This book is not to replace the standard textbooks of embryology, but to be used as a supplement in basic and advanced courses.

Each case consists of a short history detailing the most characteristic features (i.e., complications of pregnancy, parity, sex of the infant, presenting signs or symptoms), diagnosis, and treatment, followed by a discussion, generally in the form of questions and answers. This presentation encourages self-study, while further information can be acquired from the references.

The illustrations have been kept simple and are primarily diagrammatic.

The value of this book has been greatly increased by the kindness and courtesy of a number of individuals and publishers who have given us permission to make use of their photographs, x-rays, and illustrations. In addition to those acknowledged elsewhere in the text, we extend our sincere thanks to Dr. R. B. Berggren for photographs of cases 7, 8, 9, and 23; Dr. H. W. Clatworthy, Jr. for photographs of cases 5 and 24; Dr. A. P. Eaton for photograph of case 30; Dr. J. W. Kilman for photographs and x-rays of cases 12, 13, 14, and 15; Dr. P. R. Miller for photographs of cases 28, 31, 32; Dr. J. S. Morse for photographs and x-rays of cases 16, 17, 18, 19, 20, and 22; and Dr. J. P. Smith for photographs and x-rays of cases 25 and 29.

We wish to extend our sincere thanks to Mr. James R. Krentzfeld and

Miss Mitzi A. Prosser, Department of Medical Illustration, for their continued work and supervision of this project, and to their artists Marc Dockenbach, Deborah Gross, Rosalind Hagerman, Charles Hollingsworth, Marcelle Losko, Ron McLean, Steve Moon, Renate Sturtz, and Linda Tenukas for their careful interpretation of our rough sketches for the illustrations. We are greatly indebted to Mrs. Ellen Maurer and Mrs. Nell Geddie for their skill in typing, and to Miss Anita Topson for her careful reading of the manuscript.

It was a pleasure to deal with Mr. Joe Witcher and the staff of F. A. Davis Company, who showed consideration and cooperation during the preparation of the book.

Contents

TERATOLOGY

Teratology

Variation in biological structure is common. Most variations are slight and within the arbitrary limits called "normal." Teratology is the study of defective development, where these variations are not within the normal limits. Moreover, teratology encompasses two major areas: (1) congenital malformations: gross and microscopic structural defects that are present at birth and are responsible for approximately 15 per cent of the deaths in the neonatal period, and (2) congenital anomalies: physiological and metabolic disturbances and microstructural changes that damage the organism prior to birth.

INCIDENCE

At present, it is impossible to obtain accurate figures on the number of infants having congenital defects. Errors in diagnosis, lack of uniformity in collecting and analyzing data, variation in terminology, and failure to record minor defects make it difficult to compare reported incidences of congenital defects from various institutions. However, certain generalities are apparent. Excluding premature births, approximately 3 per cent of all newborn infants have a major congenital defect that is either lethal or will cause death or disability if untreated. By the end of the first year, this percentage is nearly double. In addition to fetal mortality, there are individuals who are permanently handicapped to varying degrees and individuals who are destined for a premature death owing to the presence of a major defect.

TYPES OF ABNORMAL DEVELOPMENT

Distorted Development of Body Form

Spontaneous abortions reveal many human embryos that appear as unorganized balls of cells in a chorionic vesicle. Less extreme disorganization may permit the embryo to survive longer and to show some structure characteristic of its developmental age. More difficult to understand are instances where one part of the body is radically distorted while adjacent regions appear normal (i.e., cyclopia, sirenomelus).

Growth Deficiencies and Excesses

Localized deficits in development occur in varying degrees in numerous organs. There may be agenesis of a whole organ (e.g., lung, kidney)—such failures usually are unilaterial—or specific component parts of an organ may fail to develop as in cleft palate, bifid sternum, or persistent interatrial septal defects. Developmental excesses occur in the formation of more than the normal number of parts (e.g., polydactyly, polythelism, polymastia) or in local overgrowths (e.g., premature closure of the interatrial foramen).

Malposition

These defects result from normal growth processes that occur in an abnormal location or out of time sequence (e.g., transposition of the great vessels, malrotation of the gut, annular pancreas).

Retention of Embryonic Form

In these instances, the embryonic pattern, instead of the adult form, is retained. Retention of a structure as well as its embryonic position may persist (e.g., undescended testis, pelvic kidney, imperforate anus, Meckel's diverticulum, rectovesicle fistula, hypospadias).

Metabolic and Histogenetic Disturbances

Less readily recognized are genetic and enzyme deficiencies (e.g., cri du chat syndrome, phenylketonuria) and disturbances of tissue differentiation (e.g., cretinism, gonadal dysgenesis, Hirschsprung's disease). Treatment of such disorders is an important and expanding field of pediatrics.

Many additional malformations and anomalies can be cited, but the aforementioned emphasize the range and diversity of developmental processes, disturbances of which can lead to a congenital defect. Obviously there is no simple explanation for such disturbances.

PROCESSES AND CONCEPTS IN DEVELOPMENT

To understand the modus operandi of congenital defects it is necessary to examine several developmental processes that determine the total organization of the living substance from molecules to the whole organism.

Differentiation

During the early phases of development when the number of blastomeres are rapidly increasing (i.e., growth, the process of creating more of a substance than is already present) all cells are morphologically identical and capable of accomplishing only certain basic functions such as protein synthesis and respiration. However, during formation of the germ layers, cells lose their similarity and some acquire totally new morphological characteristics and functions. The term differentiation means the creation of these new cells, tissues, or organs not previously present in development. Moreover, differentiation of a particular cell type may be dependent upon the production of unique and highly specialized enzymes. Certain general conclusions can be drawn from the differentiation process: (1) the process may involve one, a few, or many changes in form (shape, size, internal structure) and activity (cell production, proliferation rate, responsiveness, metabolism) of the cell; (2) in general, all changes are toward specialization (limitation) in the cell; (3) the change from simple to complex is relative and is actually a change at the subcellular, cellular, and supracellular levels; and (4) association of cells into groups, as in all multicellular organisms, makes possible more intercellular specialization than otherwise occurs.

Determination

Closely interlinked with differentiation is determination. Prior to the time that the cells of an early embryo become fixed they are pluripotent (i.e., can develop or act in any of several ways), but when the cells differentiate to where they can continue only in a particular direction (unipotential) they have undergone determination. In the embryo, as differentiation and determination proceed, the potency of the various cell aggregates becomes progressively restricted. Thus, the early pluripotent cells have given rise to descendants having less plasticity. The loss of plasticity can be ascertained by transplanting the cells in a controlled nondirective environment such as a tissue culture. If the cells retain their specialized structure (no dedifferentiation), differentiation has occurred and their potency is limited to further differentiation in the declared direction.

Induction

The production of a specific morphogenetic effect in the developing embryo is called induction. The tissue which elaborates the chemical substance is named the inductor or organizer while the chemical substance emitted, which determines the growth, differentiation, and specific morphogenetic response from the embryonic tissue, is termed the evocator. The ability of a tissue to react to an evocator is called competence. The pioneering work of Spemann and Mangold initiated a new field in experimental embryology. They showed that in amphibian embryos the chordamesoderm (organ field) growing inward from the dorsal lip of the blastopore had a profound effect in the molding of the body. Above the chordamesoderm the ectoderm was induced to thicken and form the neural plate, the latter then formed the neural tube. The chordamesoderm frequently is called the primary organizer because it induces formation of the neural plate, which establishes the basic topography of the embryonic body. In the formation of a complex organ a secondary organizer (e.g., the optic cup) exerts an influence on lens development and a tertiary organizer (e.g., the lens) influences the overlying ectoderm to form the cornea, etc. To be effective the organizer must be virtually in contact (surface reaction) with the responding tissue; the inference is that the evocator is a chemical substance transferred by diffusion. Furthermore, the organizer retains its capacity for induction over a limited time.

Normal differentiation of the organs involves proper occurrence of the intrinsic and relatively irreversible changes in the cells. The activity of the entire hierarchy of organizers must be properly integrated, with each exerting its inductive effect in the right place, at the right time, and in the proper strength. A large number of congenital defects cannot yet be interpreted in terms of inductor abnormalities.

ETIOLOGY OF CONGENITAL DEFECTS

A congenital defect is not necessarily the result of a *single* etiological factor, but may be influenced by heredity or genetics and/or environmental influences such as maternal medication, infection, or irradiation.

Genetic Factors

Relatively simple techniques are utilized for the examination of human chromosome structure. The normal human somatic cell contains 46 chromosomes composed of one pair of sex chromosomes and 22 pairs of autosomes. In the male, the sex chromosomes are an X and a much shorter Y chromosome. In the female, the sex chromosomes are both X's.

Mendelian Factors. Several congenital defects are explained utilizing mendelian principles. Dominant inheritance accounts for such conditions as claw hand, achondroplasia, brachydactyly and certain types of clubfoot and cleft palate. Recessive inheritance can explain albinism and phenylketonuria. Beyond such relatively simple examples are the more complex problems of polygenic inheritance, multiple alleles, and variations in gene expressivity.

Autosomal Factors. Autosomal aberrations are of two types: aberrations in number and in structure. If chromosomes are unequally divided and distributed during gametogenesis, gametes and a conceptus with greater or fewer than 22 pairs of autosomes results. This condition of having an unbalanced set of chromosomes is known as aneuploidy. When one autosome of any given pair is missing, the condition is monosomy; if an extra autosome is present, the condition is trisomy. The most common instance of trisomy is Down's syndrome (mongolism). Translocation of an autosome, giving rise to one partially deleted autosome and one autosome having additional chromosomal material, results in cri du chat syndrome.

Sex Chromosome Factors. In Turner's syndrome there are only 45 chromosomes with an XO sex chromosome complement. Such individuals are phenotypically female, but exhibit gonadal dysgenesis and variable developmental defects. Phenotypic males having an XXY sex chromosome complement exhibit Klinefelter's syndrome and characteristically are sterile. Additional defects in phenotypic males and females including sex chromosome complements of XYY, XXYY, XXXY, XXX, and XXXX are reported.

Environmental Factors

Many environmental factors and teratogens can insult the embryo in utero, but relatively few teratogenic factors are identified for man.

Infection. Viruses are implicated in many defects. If rubella (German measles) occurs during the first three months of pregnancy, there is a 10 to 20 per cent chance that the newborn infant will have defects of the eye, ear, heart, and/or brain. Influenza, smallpox, and infective hepatitis viruses also are implicated in defects. The spirochete, Treponema pallidum, can cross the placental barrier resulting in congenital syphilis; if the infection occurs early in pregnancy, abortion may result.

Drugs. Drugs, like viruses, affect the embryo during the first three months of pregnancy, the period in which organs are undergoing active development. The widespread use of thalidomide in Germany in 1960 led to a great increase in limb malformations. There is evidence that other drugs (e.g., quinine, aminopterin and antithyroids) can affect the embryo.

Mechanical. Although conclusive documentation has not established that mechanical factors can result in congenital malformations, the following possibilities are suggested: (1) intrauterine pressure on the fetus, as in oligohydramnios, may be responsible for talipes equinovarus (clubfoot); (2) an abnormally positioned limb in the developing fetus can cause pressure on the head or trunk, leading to a distortion in growth; and (3) common association of breech deliveries with congenital dislocation of the hip may indicate a lack of normal pressure forces necessary for development of an adequate acetabulum.

Nonspecific Factors. Factors such as hormones, radiation, age of the mother, dietary deficiencies of the mother, parity, season of birth, social class of the parents, geographic region, and complications of pregnancy appear to influence to some degree the incidence of certain defects.

The remaining chapters are devoted to selected cases of congenital defects: those which best serve to demonstrate basic developmental concepts.

METABOLIC AND GENETIC DISTURBANCES AND EARLY EMBRYONIC DIFFERENTIATION

1

Phenylketonuria

The patient, the second male child of unrelated parents, was delivered by elective cesarean section and weighed 2.94 kg. When the infant was 5 days of age, routine laboratory screening disclosed the urine to be positive for ferric chloride. In addition, there were increased amounts of phenylalanine, phenylpyruvic acid, and O-hydroxyphenylacetic acid in the urine. The level of phenylalanine in the blood was 42 mg. per cent (normal 2 to 4 mg.), as determined by the bacterial inhibition assay[10] and a fluorimetric procedure[18] using blood serum. Following phenylalanine loading tests, both parents were found to have less of an increase in plasma tyrosine than controls.

The infant was placed on a low phenylalanine diet (only amount needed for growth and repair, i.e., 20 mg. phenylalanine per day), and the blood level has remained below 5 mg. per cent up to the present age of 4 years and 2 months. The infant appears to have normal motor, intellectual, and behavioral development. A male sibling is normal.

DIAGNOSIS

Phenylketonuria (PKU, imbecillitas phenylpyruvica, Følling's disease, phenylpyruvic oligophrenia). Phenylketonuria is the result of an inborn error of metabolism due to a defect (absent or inactive) in the hepatic enzyme system that hydroxylates phenylalanine to form tyrosine. If undetected and untreated, the accumulation of phenylalanine in the blood and body tissues results in: (1) mental deficiency; (2) extrapyramidal signs; (3) growth retardation; (4) seizures; and (5) pigment deficiency in the hair and skin. Mental, physical, and neurological complications are usually prevented by a low phenylalanine diet begun early in life (less than two months of age); moreover, affected individuals can live a normal life span.

DISCUSSION

Garrod's recognition[6] of the relationship among gene, enzyme, and clinical abnormality twenty-five years before the first cases of PKU were described by Følling[5] provided the conceptual basis for the disease. Penrose[22] suggested that PKU might be attributed to recessive genes, while Jervis[13] correctly proposed that substitution of a single gene was the cause. That same year, 1937, the name phenylketonuria was introduced.[23]

Hereditary Factors

What is the hereditary basis for phenylketonuria? Phenylketonuria results from a single autosomal recessive gene, two of which, when combined in an offspring, will produce the disorder. In an infant having PKU, both parents possess the defective form of one of the two genes controlling phenylalanine hydroxylase. On the average, one of four offspring from the heterozygous parents have both defective genes; in these infants there is no hydroxylation of phenylalanine and PKU results. In the United States both sexes are affected in approximately equal numbers and PKU occurs in 1/10,000 live births.[19]

Clinical Features

What is the characteristic picture of an infant having untreated PKU? The infant appears clinically normal at birth; however, within the first ten months it is apparent that the usual developmental milestones, i.e., sitting, walking, talking, are delayed, with no significant loss after the third year. The child's behavior is restless, jerky, and fearful with muscular hypertonicity, hyperactive tendon reflexes, and adventitious hyperkinesis. Apart from the central nervous system, developmental abnormalities only of the skin (eczema) and teeth and bones (microcephaly)—all of which have an ectodermal origin—are reported.

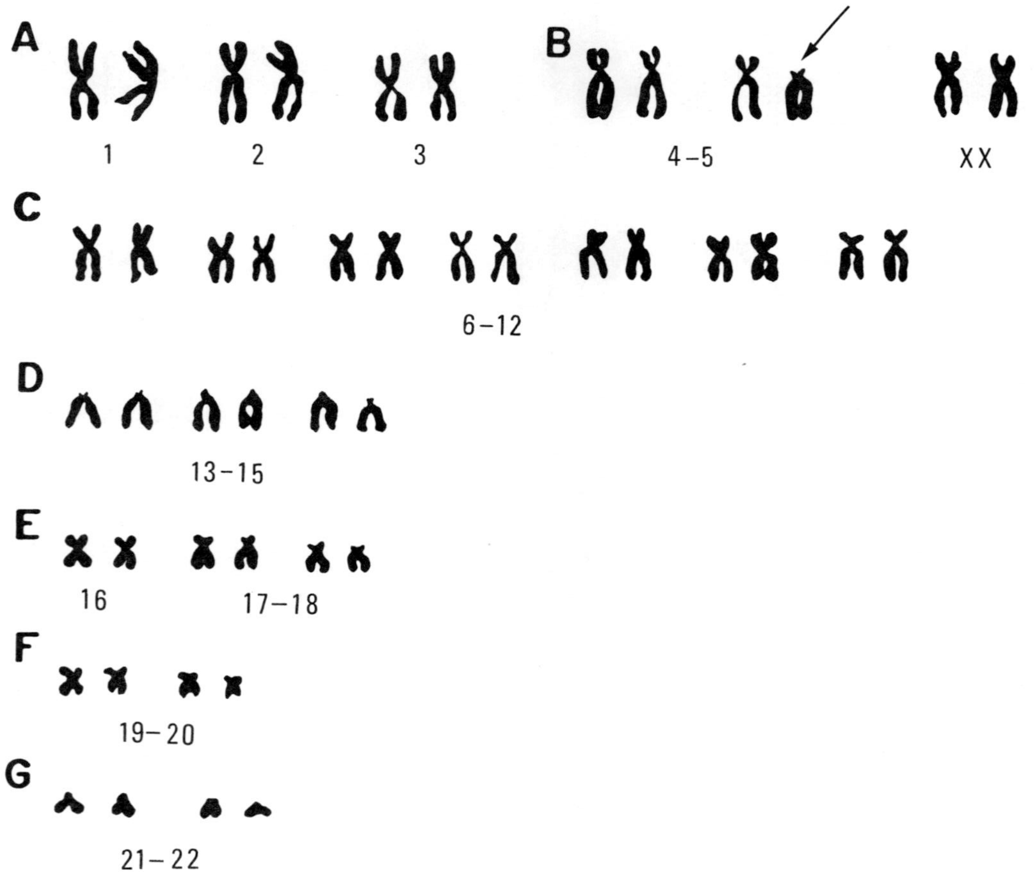

Karyotype of the *cri du chat* syndrome showing deletion of the short arm of chromosome B5, *arrow*.

2

Partial Deletion of the Short Arm of Chromosome 5 (5p-)

A gravida II, para II mother in the fortieth week of gestation gave birth to a female infant having a weight of 2.5 kg. (5 lbs., 10 oz.), low for this week of gestation. The mother's age was 32 years at conception; that of the father, 40 years. The family history revealed no consanguinity. The infant's face was round, with the head small in relation to the facial bones. Micrognathia, epicanthus, hypertelorism, antimongoloid slant, low-set ears, and a short neck, as well as severe mental and physical retardation, were apparent. Her cry was weak and plaintive, resembling the mewing of a cat. She did not support her head until 9 months of age, and at 4 years she still does not walk.

FINDINGS

Routine Laboratory Studies

Considered normal and not contributory to the diagnosis.

Dermatoglyphics

Revealed a transverse palmar crease and distal axial triradius.

Cytogenetic Findings

Chromosomal preparations of the infant and parents were made, using the peripheral blood leukocyte culture technique. The parents revealed no chromosomal abnormalities, but the infant demonstrated a karyotype containing a chromosome of the B4-5 group having partial deletion of the short arm. Chromosomal autoradiography revealed the abnormal chromosome to be one of the pair of the B4-5 group which terminates DNA replication early (B5). No other chromosomal abnormalities were found.

DIAGNOSIS

Cri du chat syndrome.

DISCUSSION

Jerome Lejeune and his colleagues (1963)[14] described the first three cases of a new syndrome that is characterized by a weak mewing cry during infancy, giving rise to the descriptive name "cri du chat." They showed this syndrome to be caused by partial deletion of the short arm of a B group chromosome, later identified by DNA radioautographic labeling studies[7] as chromosome number 5.

Incidence

In the general population, the incidence of cri du chat syndrome may range from 7/744 to 1/2,159.[20] Parental age, order of birth, and consanguinity apparently are of no significance. At present, no specific gene locus has been mapped on the deleted segment (5p-); hence, genetic marker studies offer little help in resolving this problem.

Etiology

Partial deletion of the short arm of chromosome number 5 appears as a new aberration in the vast majority of cases, and therefore as sporadic occurrences within families. Two possibilities are considered to explain this situation: (1) mosaicism: the presence of cells in a single individual that are genetically different despite their common origin, and (2) translocation: the fragmentation of heterologous chromosomes in such a way that the fragmented portion of one chromosome becomes attached to the broken end of another chromosome, thereby increasing its length.

The parents of one patient of Lejeune and associates[16] demonstrated that the phenotypically normal mother was a balanced translocation carrier involving members of the B4-5 and D13-15 groups. The affected child apparently received the deleted B5 chromosome but not the member of the D13-15 group to which the deleted material was attached.

Associated Malformations

Occasionally congenital defects occur with this syndrome, e.g., heart disease (variable in type), cleft lip and cleft palate, preauricular skin tag, facial asymmetry, bifid uvula, short neck, short metacarpals, clinodactyly, inguinal hernia, cryptorchidism, hemivertebra, and absent kidney or spleen.

Natural History

The majority of reported cases are in the young; thus, it is difficult to depict the natural history of the disease at present. The mewing cry, ascribed to abnormal laryngeal development (during phonation and crying, the voice is weakened by air leakage because of failure of the cords to approximate posteriorly, and the high-pitched mewing cry results from the approximated anterior portions of the cords), becomes less pronounced with age, making the diagnosis increasingly difficult in older patients.

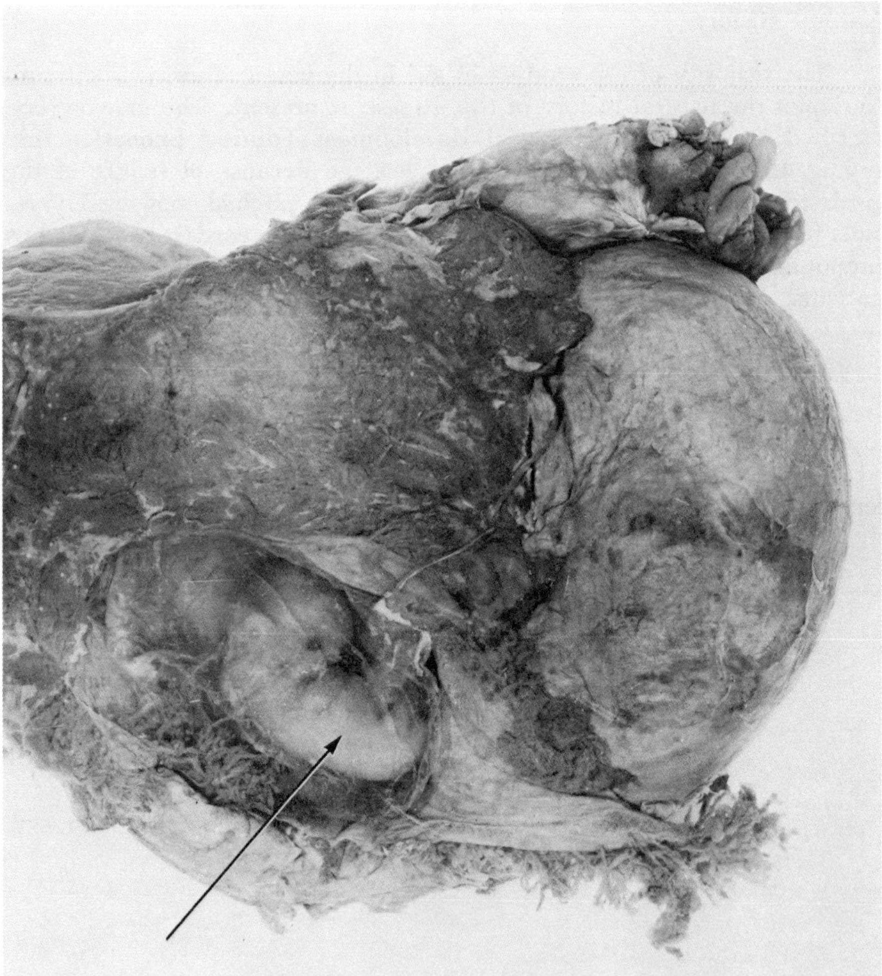

Eight-week-old tubal pregnancy, *arrow*.

3

Ectopic Pregnancy

A 30-year-old Negro female, gravida VI, para IV, with two previous abortions, was admitted to the hospital complaining of increasingly severe lower abdominal pain associated with syncope. Slight vaginal bleeding, which had begun two days previously, was accompanied by nausea but no vomiting. The last normal menses was eight weeks prior to admission.

Physical examination revealed a moderately distended and diffusely tender abdomen. Upon manipulation, tenderness was also present in the cervix, uterus, and right and left adnexa. Bowel sounds were active. A serous discharge was found upon vaginal examination. A culdocentesis aspirated blood from the cul-de-sac (pouch of Douglas). The pregnancy test was negative.

A laparotomy was performed. The abdominal cavity contained 500 ml. of clotted and unclotted blood. The left ovary had a corpus luteum of pregnancy. The ampulla of the right uterine tube was a bulbous hemorrhagic mass showing active bleeding. A right salpingectomy was performed and the clots and free blood evacuated. The postoperative course was uneventful and the patient was discharged on the fifth day.

The pathology report follows:

1. Gross description: The distal 8 cm. of the right uterine tube is enlarged, with a patent fimbriated end. There is one small, smooth-lined, clear-filled paratubal cyst. The opened cyst contains a fetus having a crown rump (CR) length of 2 cm.
2. Microscopic description:

   A. Proximal portion right uterine tube. The lumen is patent. A scattering of inflammatory cells are within the muscularis.

B. Distal portion right uterine tube and placenta. Immature chorionic villi and blood clots are present. The blood clots are between the placenta and the epithelial lining of the tube.

DIAGNOSIS

Right tubal pregnancy.

DISCUSSION

Ectopic pregnancy occurs on the average in 5 of 1,000 deliveries[15] with variations between different races and socioeconomic groups. Ectopic pregnancy was first described by Albucasis[24] in the eleventh century, while in 1604 Riolanus the younger[12] detailed one of the earliest cases of tubal gestation with classical symptomatology of tubal rupture. In 1699 Mauriceau and Vassal[21] described a pregnancy in a rudimentary uterine horn, and in 1883 Sänger[25] performed the first amputation of a rudimentary uterine horn containing a pregnancy.

Embryology

Describe the development of the ovum from ovulation through

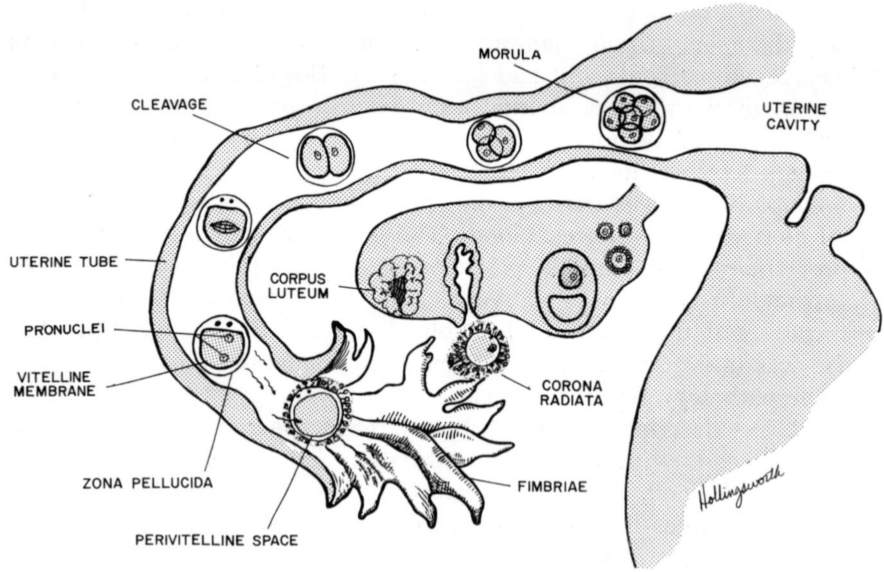

Fig. 1 Stages in the passage of the ovum into the uterine tube, its fertilization, and its subsequent development.

fertilization and implantation as a tubal pregnancy. At ovulation (Fig. 1) the secondary oocyte, with an eccentrically placed second polar spindle, is surrounded by a plasma (vitelline) membrane, outside of which lies the first polar body. This secondary oocyte and first polar body are within the zona pellucida; the latter is surrounded by the corona radiata. Upon ovulation the oocyte is swept into the fimbriated end of the uterine tube, which is closely applied to and partially covers the ovary.

Fertilization normally occurs in the outer third of the uterine tube. As sperms approach the oocyte, their acrosomal caps dissolve, releasing hyaluronidase, which causes a separation of the cells composing the corona radiata and facilitating penetration of the zona pellucida. The sperm adhere to the oocyte through an antigen-antibody-type reaction that occurs between fertilizins at the surface of the oocyte and antifertilizins in the acrosomal cap. Subsequently, the oocyte is activated and forms a fertilization membrane around the fluid-filled perivitelline space. The fertilization membrane prevents entry of additional sperms. The second polar spindle now completes division, with the formation and extrusion of the second polar body and the formation of the female pronucleus. The head of the sperm forms the male pronucleus. The nuclear membranes of the female and male pronuclei disappear, and with their centrioles the pronuclei migrate towards the center of the cell where their chromosomes become arranged around the equator just prior to the first cleavage division. Thus, fertilization results in: (1) restoration of the diploid number of chromosomes; (2) determination of the chromosomal sex of the individual; and (3) the initiation of cleavage.

Cleavage is a rapid succession of mitotic divisions resulting in the production of an increased number of smaller cells, called blastomeres. The principal consequences of cleavage are: (1) division of the protoplasm of the zygote amongst the blastomeres; and (2) increased mobility of this protoplasm. In addition, the fertilized ovum is considerably larger than the average adult cell, and cleavage results in an approximation in size of the developing cells to that size characteristic of the definitive cells of the organism. During this initial period of cleavage, certain unknown factors influence the implantation and further development of the embryo in an abnormal position, e.g., tubal.

Etiology

What factors may contribute to the formation of ectopic, particularly tubal, pregnancies? Factors that delay or prevent tubal transport of the fertilized ovum, such as (1) gonorrhea; (2) postabortion; (3) puerperal infection; (4) hormonal influence; (5) conceptus itself; (6) congenital abnormalities; and (7) chronic inflammatory disease, or anything that impairs tubal function.

Grapelike appearance of a hydatidiform mole.

4

Trophoblastic Disease

A 38-year-old gravida III, para II female was admitted to University Hospital in May, 1970, complaining of excessive vaginal bleeding of less than one day's duration. Her last menstural period (LMP) was January, 1970. Prior to this time she had menstruated regularly. Gynecologic examination in April, 1970, revealed a uterus corresponding to a normal three-month pregnancy. Examination at the present admission disclosed an intrapelvic and intra-abdominal mass located in the midline and corresponding to a uterine enlargement of a six-month pregnancy (an apparent uterine growth of three months' pregnancy occurring within one month's time). The cervix was soft, and the os cervix was dilated.

DIAGNOSIS

Missed abortion, hydatidiform mole, or multiple pregnancy?

TREATMENT

Bioassays revealed HCG excretion above 50,000 I.U./24 hours (a peak excretion of 7,000 to 500,000 I.U./24 hours normally occurs between the tenth and fourteenth week; if after the sixteenth week, hydatidiform mole is probable). Fetal ECG verified the absence of fetal heart sounds; spontaneous expulsion of a small portion of tissue with a grapelike appearance, combined with ultrasound, confirmed the diagnosis of hydatidiform mole (a cystic, hydropic swelling of the chorionic villi, accompanied by variable hyperplastic and anaplastic changes in the chorionic epithelium with no apparent embryo). Curettage of the uterine cavity was performed to remove the remainder of the mole. Close examination of the mole revealed a small chorionic sac; the sac was opened and found to be empty.

Quantitative determination of HCG was done weekly until results were normal for three consecutive weeks. The determination was repeated monthly for one year, during which time pregnancy was avoided. Simultaneously, chest x-rays to detect lung metastases were made. Two years postoperatively the patient was well, the lung fields were normal as determined by x-rays studies, and there was no evidence of increased HCG.

DISCUSSION

In the United States, hydatidiform mole (molar pregnancy) occurs in 1/1,500 pregnancies.[8] Although Aetius wrote about hydatidiform mole in the sixth century, von Grafensberg (1565) is credited with its classical description.[2] Gieser (1847)[2] was the first to attempt a correlation between pregnancy and hydatidiform mole and considered the latter to be a result of changes in pathologic ova. Sänger (1889)[26] suggested that hydatidiform mole was a specialized tumor derived from the decidua of pregnancy, while Marchand (1895)[17] demonstrated that these tumors invariably were the sequel of normal pregnancy, abortion, and hydatidiform mole. Despite the non-neoplastic nature of the hydatidiform mole, its potential malignancy (choriocarcinoma) is recognized, and is approximately parallel to the degree of hyperplasia and anaplasia of the trophoblast.

Embryology

Discuss normal placentation. The zygote (Fig. 2), composed of a double-layered ball of cells (morula), surrounded by the zona pellucida, enters the uterine cavity on approximately the sixth day following ovulation. At this time the endometrium is in the luteal phase and the uterine cavity is lined by mucus. The morula is suspended in this mucus for two to three days, during which time it develops into a spherical blastocyst by the following modifications: (1) the zona pellucida disappears; (2) nutrition (embryotrophe) occurs via fluid passing from the uterine cavity into the intercellular spaces between the double layer of cells composing the blastocyst (at this time the inner cell mass has formed from the inner layer of the blastocyst); (3) the intercellular spaces become confluent as a result of the increased amount of fluid, forming a single cavity, the blastocyst; and (4) the inner cell mass that forms

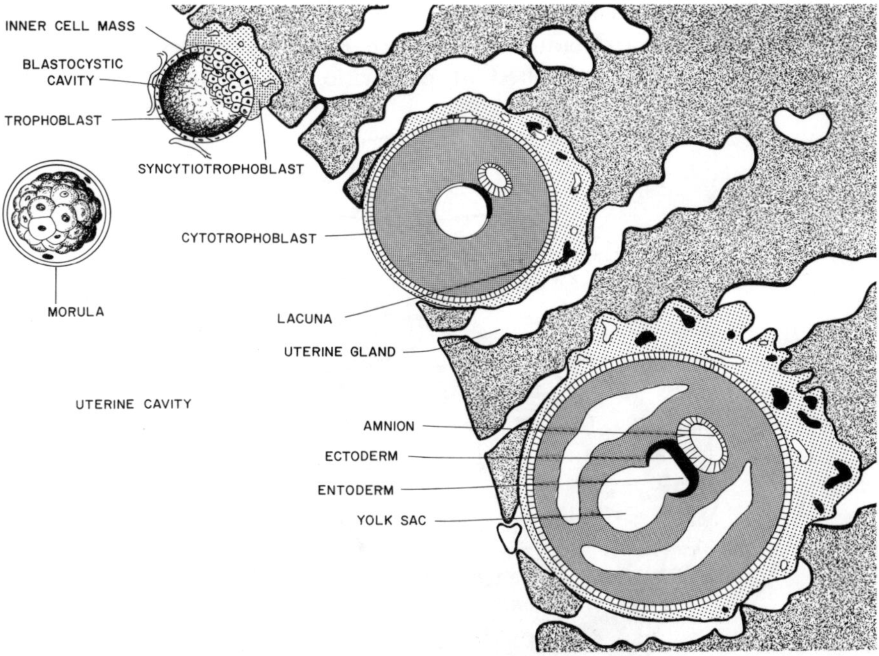

Fig. 2 Cleavage and implantation. (From William, P. L., and Wendell-Smith, C. P.: Basic Human Embryology. Pitman Medical Publishing Co., Ltd., London, 1966, with permission.)

the future embryo and some extraembryonic membranes attaches eccentrically to the inner aspect of the outer layer of cells of the blastocyst. The outer layer is now termed the trophoblast (Fig. 2).

One week after ovulation, the trophoblastic cells attach to the uterine epithelium by erosion and penetration of the underlying endometrial stroma. The cells composing the trophoblast differentiate into two types: (1) the cytotrophoblast, an inner layer adjacent to the developing embryo; and (2) the syncytiotrophoblast (syncytium), an outer layer adjacent to the uterine epithelium. The above two cell layers will form the chorion, and by differentiation the latter gives rise to the chorionic plate.

Nine to eleven days (Fig. 2) after ovulation, isolated vacuoles appear within the syncytium. Subsequently, these vacuoles become confluent, forming lacunae which are the beginning of the future intervillous space of the placenta.

The uterine wall is composed of an endometrial, a myometrial, and a serosal layer. The endometrium (Fig. 3) itself is composed of: (1) a zona compacta; (2) a zona spongiosa; and (3) a zona basalis. The zona compacta and the zona spongiosa constitute the decidua. The decidua undergoes a decidual reaction in response to the implantation of the trophoblast and is shed at parturition. The decidua (Fig. 3)

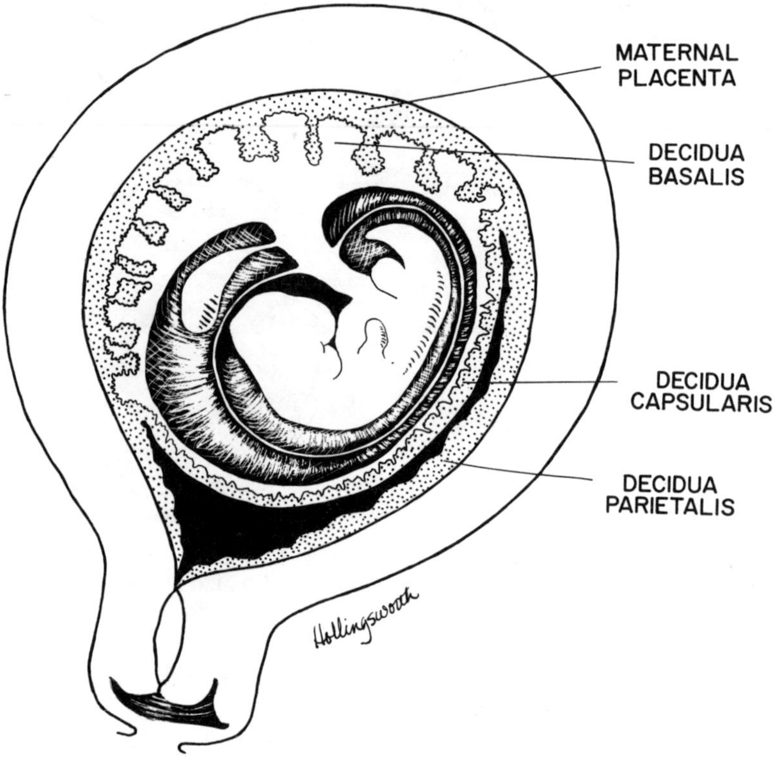

Fig. 3 Embryo in utero at 5 weeks.

consists of three topographical regions up to the fourth month of gestation: (1) the decidua basalis; (2) the decidua capsularis; and (3) the decidua parietalis. Twelve days following ovulation, the capillaries of the decidua basalis become congested and dilated and form interconnecting sinusoids. The syncytium implants into the decidua, and maternal blood flows from the sinusoids to the developing lacunae (histotrophic nutrition). As implantation continues, larger maternal vessels are tapped and the syncytium becomes directly continuous with the endothelial lining of these vessels. By the fifth month, the uterine cavity is obliterated by fusion of the decidua capsularis with the decidua parietalis. Subsequently, the decidua capsularis thins and disappears, leaving the decidua basalis to form the maternal portion of the placenta.

Chorionic villi, the essential structure of the definitive placenta, begin to form at the third week of gestation (Fig. 4A). The cytotrophoblast extends into the syncytium to form primary villi as the first step in the development of chorionic villi. As extension occurs, a core of mesodermal tissue extends into the primary villi, which are now known

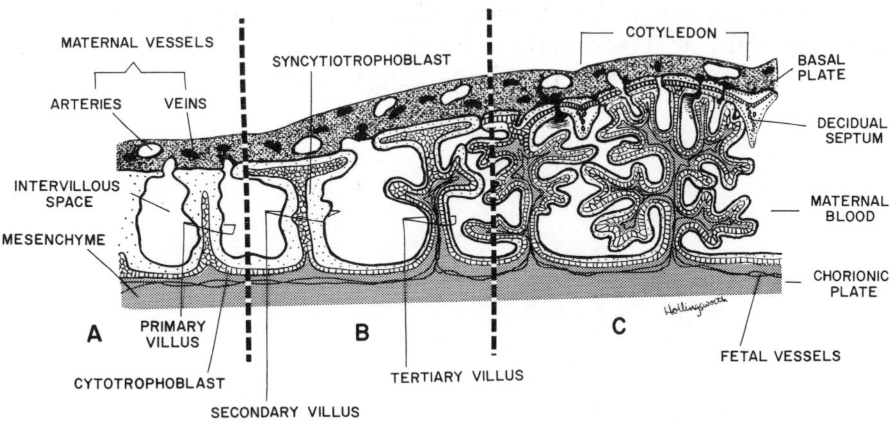

Fig. 4 Enlargement of placental site in Figures 2 and 3; *A*, Third to fifth week of gestation; *B*, Second month of gestation; C, Middle of second trimester.

as secondary villi. These villi become united distally to form a chorionic shell that is penetrated by maternal vessels. This penetration aids in anchoring fetal to maternal portions of the placenta.

In the fourth to fifth week of gestation, the secondary villi are penetrated by blood vessels and are now called tertiary villi (hemotrophic nutrition). These villi will branch and fuse with one another (Fig. 4B).

By the fourth prenatal month, the cytotrophoblastic cells begin to retrogress, probably the result of a reduction in mitotic rate, leaving villi covered predominantly by syncytium. Concurrently septa from the decidua basalis extend into the intervillous spaces dividing the placenta into 15 to 20 units called cotyledons (Fig. 4C). The septa do not extend to the chorionic plate, thus communications between cotyledons remain. The placenta rapidly decreases in growth toward the end of the fourth month.

Cause

Discuss the development of hydatidiform mole. Formation of hydatidiform mole begins between the third and fifth week of development at the time the fetochorionic circulation normally becomes established. However, no chorionic circulation develops, since the fetus is either absent, defective, or dead, and dissolution of the mesodermal core within the villi results. Concomitantly, the trophoblast, nourished by maternal blood within the intervillous spaces, survives and secretes and transfers products from this maternal blood into the chorionic villi. Since there is no chorionic circulation the liquid products accumulate, distending and converting the villi into cysts. The chorionic villi, in spite of the defective intrinsic blood supply, continue to grow rapidly within the uterine cavity, giving rise to a hydatidiform mole.

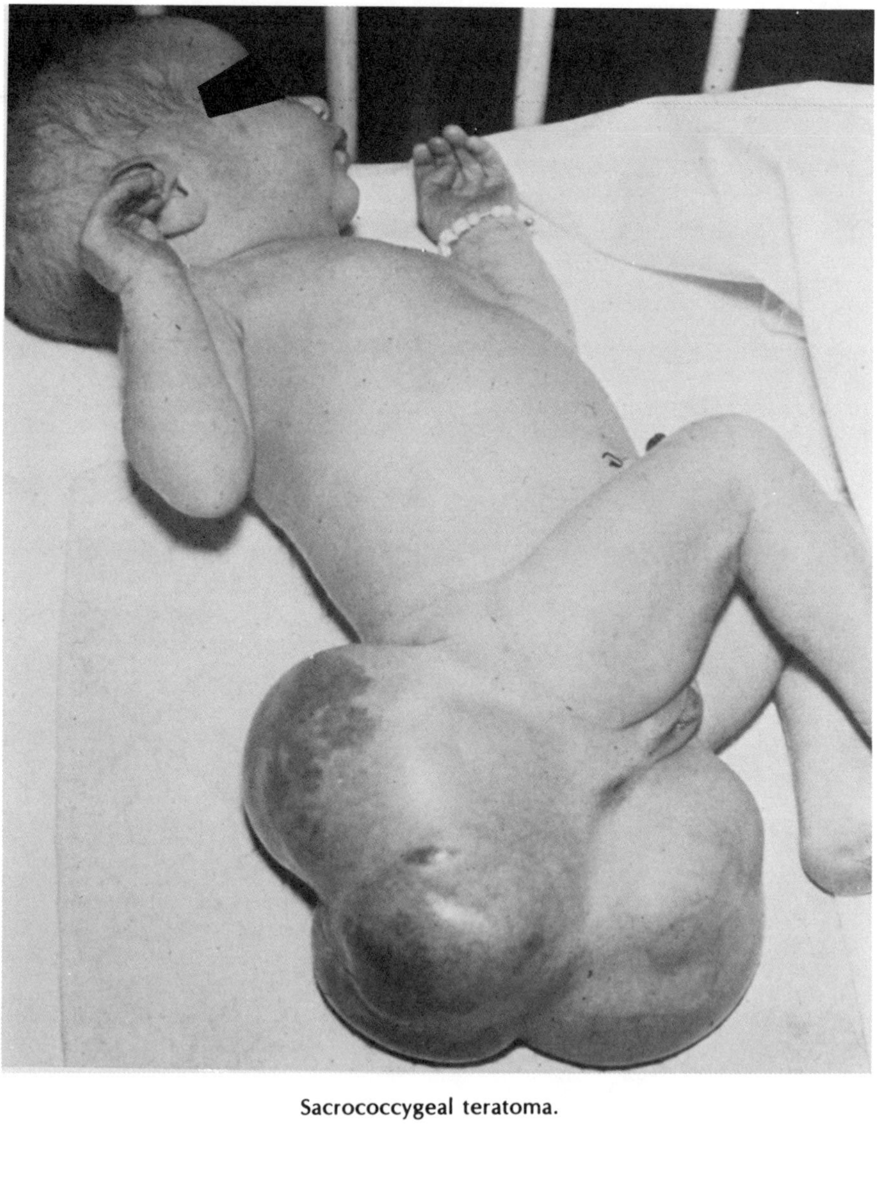

Sacrococcygeal teratoma.

5

Teratoma

A 29-year-old woman previously having had four normal deliveries was in labor and admitted to the emergency room. The infant's head, upper limbs, and trunk were delivered normally, at which time labor became obstructed by the buttocks and lower limbs. Traction failed to complete delivery and a cesarean section was performed and the infant extracted.

EXAMINATION

The female infant weighed 3,600 gm. and appeared active and well at birth. A raised skin-covered mass devoid of hair and 20 cm. in diameter protruded from the coccygeal region, slightly displaced to the left of the midline. X-ray studies of the pelvis revealed a soft tissue mass in the gluteal area and no calcification of the coccygeal vertebra. Pelvic extension of the mass could not be determined and surgery was performed at age 48 hours.

DIAGNOSIS

Provisional diagnosis: benign sacrococcygeal teratoma.

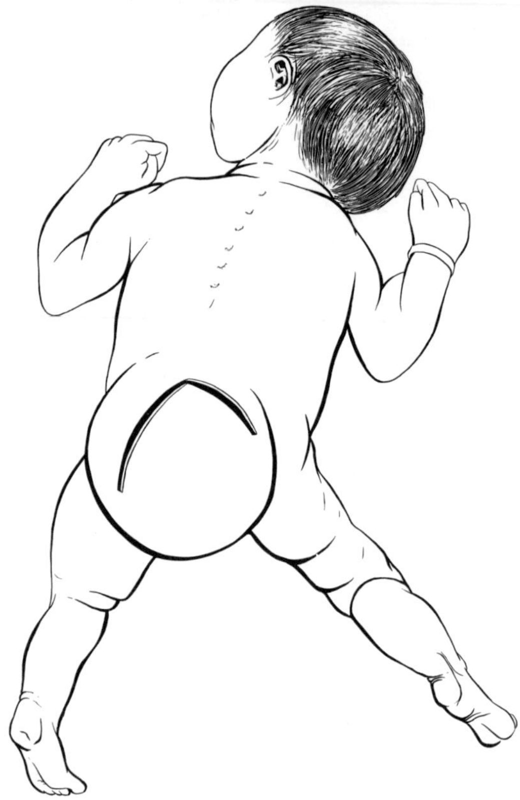

Fig. 5 Removal of a sacrococcygeal teratoma via an inverted V-shaped incision.

TREATMENT

The teratoma, an encapsulated tumor beneath the subcutaneous tissues, was excised through an inverted V incision (Fig. 5) over the lower sacrum and buttocks. Prior to removal of the tumor, the middle sacral artery and vein and the lateral sacral vessels supplying it were ligated. The tumor was separated from the rectum and the coccyx excised and left attached to the tumor, after which the wound was closed.

Histological sections of the tumor showed cysts lined by glandular epithelium; fat, muscle, and connective tissue; acini resembling the pancreas; cartilaginous material, sebaceous glands; hair follicles and sweat glands, which confirmed the provisional diagnosis of benign sacrococcygeal teratoma. The postoperative course was uneventful, and the infant was discharged at 8 days of age. At 15 months after surgery the female infant is healthy and well developed with no evidence of tumor recurrence.

DISCUSSION

The earliest written record of sacrococcygeal teratoma is found on a cuneiform tablet of Babylonian origin from Chaldea around 2000 B.C.[1] The first successful surgical extirpation of a sacral teratoma was reported by Blizard in 1841.[27] Descriptive and speculative investigation of teratomas has led to a vast accumulation of literature.[3, 4, 9]

Definition of Teratoma

Give a generalized definition or description of teratoma that includes all clinical manifestations. Teratomas occur primarily in the female and are present in 1/35,000 live births.[11] Teratomas are tumors arising from pluripotential embryonic cells; these cells have escaped the influence of an organizer and occupy a preaxial median or a closely paramedian position. The tumors usually arise from more than one germ layer and form a variety of tissues foreign to the area in which they develop. These tumors generally are of two types: (1) a benign cystic or solid mass which grows at the same rate as the embryo, fetus, or infant and is asymptomatic only because of its size or by its displacement of adjacent organs, or (2) a rapidly growing, invasive or metastatic mass, generally arising from one layer. In teratomas of the newborn (congenital), any of the involved tissues may become malignant.

Embryology

From the generalized definition of a teratoma, suggest how these tumors might develop. These benign or malignant tumors may arise from detached embryonic cell nests (e.g., neuroenteric canal, postanal gut, proctodeal membrane, anal plate) or from embryonic misplacement of tissues, which retain some or all of their totipotentiality (capable of forming a new individual) and for reasons unknown grow, despite an apparent lack of the required organizer. The median or paramedian location suggests an origin from embryonic cells of the primitive streak or primitive knot.

In general, discuss briefly what the primitive streak and the primitive knot are. Discuss formation of the intraembryonic mesoderm and the notochordal process. The primitive streak is an ectodermal area at the caudal edge of the bilaminar embryo, and early in the third week

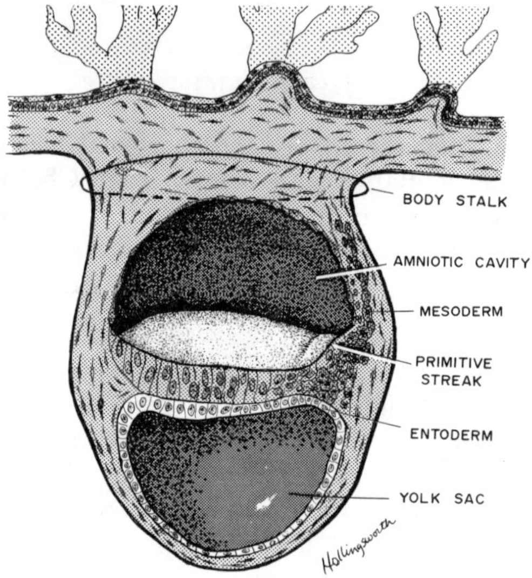

A

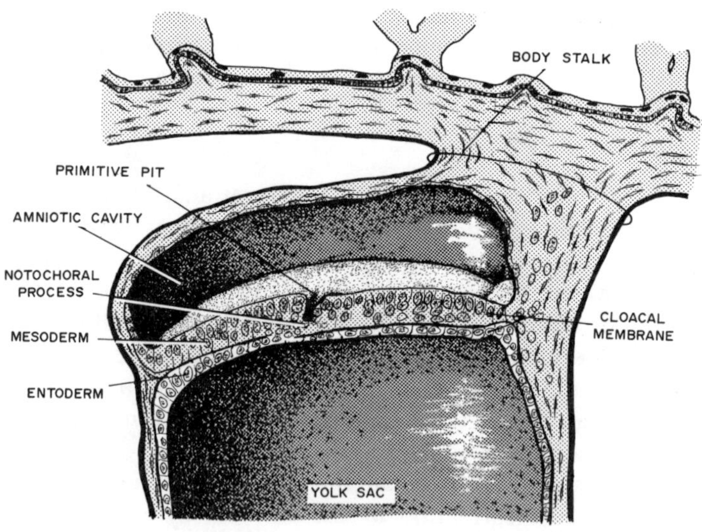

B

Fig. 6 Development of the primitive streak: *A,* Sagittal section, 2.5-week-old embryo; *B,* Sagittal section, 3-week-old embryo; *C,* Dorsal view, 4-week-old embryo.

of gestation (Fig 6A) the primitive streak appears as a proliferation of cells that bulge into the amniotic cavity in this caudal region. Cells migrate laterally from the streak between the ectoderm and entoderm to form the third germ layer of the embryo, the intraembryonic mesoderm. A further thickening of ectoderm at the cephalic end of the streak forms the primitive knot (Hensen's node), in the center of which occurs an inward movement of surface cells resulting in the development of the primitive pit (blastopore) (Fig. 6B). Surface cells migrate through the blastopore between the ectoderm and entoderm forming the notochordal or head process. As the embryo increases in length, the primitive streak and knot are carried caudally and persist until the end of the somite stage (5 mm), when they undergo retrogressive change and rapidly diminish (Fig. 6C).

Discuss briefly how germinal and somatic cells might give rise to a teratoma. In the early stages of development and segmentation of the zygote, the first blastomeres are totipotent; later these same blastomeres are multipotent (capable of forming all three germ layers). These multipotent blastomeres give rise to two definitive cell types: (1) germinal cells, which normally develop into spermatozoa or ova, and (2) somatic cells, which develop into the remaining body cells proper. If either germinal or somatic cell is detached and stimulated to grow outside of its normal location, a teratoma may arise.

Specificity of the Germ Layers

What is the germ layer theory of specificity? Does the histological

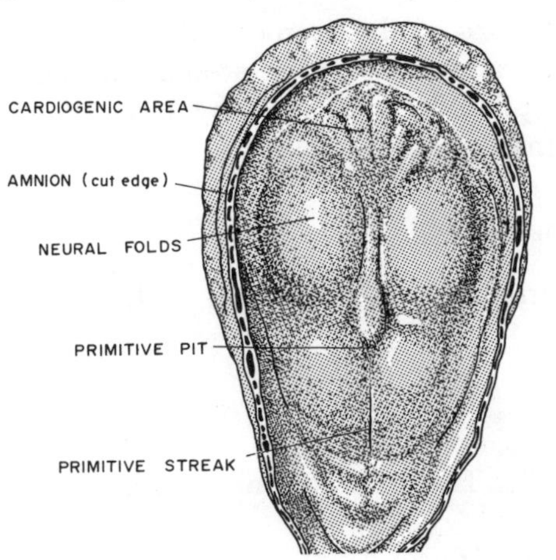

C

Fig. 6 (continued. See legend on facing page.)

appearance of a teratoma help to substantiate or refute this theory? The cells of an embryo prior to gastrulation are capable of differentiating into widely diverse tissues under appropriate circumstances. After gastrulation this plasticity is progressively lost, each tissue becoming more restricted to one of the three germ layers (ectoderm, mesoderm, entoderm). The developmental potential of the germ layers is of two types: (1) actual, what the germ layers will form under normal conditions, and (2) total, what the germ layers are capable of forming under normal, but diversified, conditions or under experimental influences. A teratoma represents the area of total potential of the germ layers and supports the theory of specificity, because teratomas frequently show associations of structures occurring under normal conditions, e.g., cartilage with respiratory epithelium, bone with teeth, sebaceous glands with hair.

Absolute adherence to the germ layer theory is unwise, due to experimental findings regarding the origin of certain structures, e.g., the musculature of the iris, branchial cartilages, meninges, and dermal pigment cells.

Sites of Teratomas

Teratomas are found in diverse areas of the body, e.g., pituitary, pineal, face, thyroid, mediastinum, stomach, gonads, kidney, and retroperitoneal.

Differential Diagnosis

What malformations of the sacrococcygeal region may be confused with a teratoma? A sacrococcygeal teratoma may be misdiagnosed as a perirectal abscess, pilonidal cyst, meningocele, chordoma, neurenteric cyst, or hemangioma.

Associated Malformations

A random distribution of congenital malformations are associated with teratomas.

REFERENCES

1. Ballantyne, J. W.: The teratological records of Chaldea. Teratologia 1:127, 1894.
2. Bardawil, W. A., Hertig, A. T., and Velardo, J. T.: Regression of trophoblast. I. Hydatidiform mole; a case of unusual features,

possibly metastasis and regression; review of literature. Obstet. Gynec. 10:614, 1957.
3. Berry, C. L., Keeling, J., and Hilton, C.: Teratomata in infancy and childhood—a review of 91 cases. J. Path. 98:241, 1969.
4. Donnellan, W. A., and Swenson, O.: Benign and malignant sacrococcygeal teratomas. Surgery 64:834, 1968.
5. Følling, A.: Über Ausscheidung von Phenylbrenztraubensäure in dem Harn als Stoffwechselanomalie in Verbindung mit Imbezillität. Hoppe Seyler Z. Physiol. Chem. 227:169, 1934.
6. Garrod, A. E.: Inborn Errors of Metabolism. Oxford, London, 1909.
7. German, J., Lejeune, J., Macintyre, M. N., and de Grouchy, J.: Chromosomal autoradiography in the Cri du Chat syndrome. Cytogenetics 3:347, 1964.
8. Goldstein, D. P., and Reid, D. E.: Recent development in the management of molar pregnancy. Clin. Obstet. Gynec. 10:313, 1967.
9. Gross, R. E., Clatworthy, H. W., Jr., and Meeker, I. A., Jr.: Sacrococcygeal teratomas in infants and children: A report of 40 cases. Surg. Gynec. Obstet. 92:341, 1951.
10. Guthrie, R., and Susi, A.: A simple phenylalanine method for detecting phenylketonuria in large populations of newborn infants. Pediatrics 32:338, 1963.
11. Hunt, P., Van Leeuwen, G., Bingham, H., and Sights, W.: Sacrococcygeal teratomas. A report of three cases and survey of present knowledge. Clin. Pediat. 7:165, 1968.
12. Jarcho, J.: Ectopic pregnancy. I. With special reference to abdominal pregnancy. Am. J. Surg. 77:273, 1949.
13. Jervis, G. A.: Phenylpyruvic oligophrenia. Arch. Neurol. Psychiat. 228:944, 1937.
14. Lejeune, J., Lafourcade, J., Berger, R., Vialette, J., Boeswillwald, M., Seringe, P., and Turpin, R.: Trois cas de deletion partielle des bras courts d'un chromosome 5. C.R. Acad. Sci. (Paris) 257:3098, 1963.
15. Lihfeldt, H., Tietze, C., and Gorstein, F.: Ovarian pregnancy and the intrauterine device. Am. J. Obstet. Gynec. 108:1005, 1970.
16. Macintyre, M. N., Staples, W. I., LaPolla, J. J., and Hempel, J. M.: The "cat cry" syndrome. Am. J. Dis. Child. 108:538, 1964.
17. Marchand, F.: Ueber die sogenannten decidualen Geschwülste im Anschluss an normale Geburt, Abort, Blasenmole und Extrauterinschwangerschaft. Mchr. Geburtsh. Gynaek. (Berlin) 1:419, 1895.

18. McCaman, M. W., and Robins, E.: Fluorimetric method for the determination of phenylalanine in serum. J. Lab. Clin. Med. 59:885, 1962.
19. Menkes, J. H.: The pathogenesis of mental retardation in phenylketonuria and other inborn errors of amino acid metabolism. Pediatrics 39:297, 1967.
20. Miller, O. J., Breg, W. R., Warburton, D., Miller, D. A., de Capoa, A., Allderdice, P. W., Davis, J., Klinger, H. P., McGilvray, E., and Allen, F. H., Jr.: Partial deletion of the short arm of chromosome No. 4(4p-): Clinical studies in five unrelated patients. J. Pediat. 77:792, 1970.
21. Mulsow, F. W.: Pregnancy in rudimentary horn or uterus. Am. J. Obstet. Gynec. 49:773, 1945.
22. Penrose, L. S.: Inheritance of phenylpyruvic amentia. Lancet 229:192, 1935.
23. Penrose, L. S., and Quastel, J. H.: Metabolic studies in phenylketonuria. Biochem. J. 31:266, 1937.
24. Ricci, J. V.: The Genealogy of Gynaecology. The Blakistan Company, Philadelphia, 1943.
25. Sänger, M.: Ueber Schwangerschaft im rudimentaren Nebenhorn bei Uterus duplex. Zbl. Gynaek. 7:324, 1883.
26. Sänger, M.: Zwei aussergewöhnliche Fälle von Abortus. Zbl. Gynaek. 13:132, 1889.
27. Standley, E.: On congenital tumors of the pelvis. Med. Chir. Trans. Lond. Ser. 2, 6:235, 1841.

HEAD AND NECK

6

Dacryostenosis

A 3-week-old female infant is brought to the pediatrician because of a sticky white material constantly present in the inner corner of the left eye and because of a continuous overflow of tears (epiphora) on the same side. Initial examination of the eye reveals a normal conjunctival surface with no inflammation; there is puffiness of the eyelids and a swelling in the area of the lacrimal sac. Slight pressure on the swelling releases a mucopurulent discharge into the lacrimal lake of the eye, indicating an obstruction of the lacrimal duct. Downward massage over the lacrimal sac fails to relieve the obstruction. Conservative therapy of massage of the lacrimal sac is performed several times a day. Antibiotic drops, to alleviate any possible inflammation or infection, are applied daily. At 4 months the child is asymptomatic; tearing and mucous discharge have subsided, indicating patency of the lacrimal duct, mitigating the need for probing or irrigation. No symptoms are present at 2 years.

DIAGNOSIS

Congenital impatency of the nasolacrimal duct on the left side due to the persistence of the nasolacrimal membrane.

DISCUSSION

Born (1876)[5] was the first investigator to correctly interpret the early stages of development of the nasolacrimal passage, while Schaeffer's (1912)[37] description of the genesis and development of this passage in human specimens, using 30-day embryos to full-term fetuses, was the most accurate.

Investigators[35, 38] have established structural abnormalities and delayed development as the causes of nasolacrimal obstruction. In approximately 73 per cent[33, 45] of newborn infants, a nasolacrimal passage is impatent due to the persistence of the nasolacrimal membrane. Recent studies[24, 33, 34] have concentrated on the treatment and alleviation of this obstruction.

Embryology

Explain embryologically the cause(s) of epiphora. In the five-week-old embryo (Fig. 11B), the lacrimal passage begins to develop from an epithelial cord of cells along the floor of the naso-optic furrow, an area between the lateral nasal process and the maxillary process.

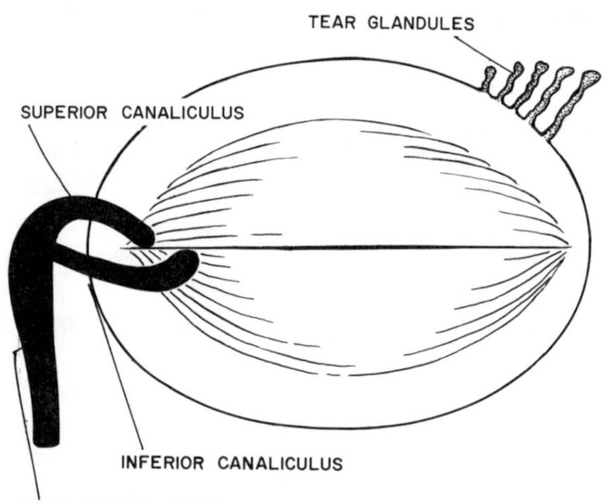

Fig. 7 Developing lacrimal apparatus, 6-week-old embryo.

In the sixth week of development (Fig. 7), the epithelial cord becomes detached from the surface ectoderm and is buried in the mesenchyme. The cord grows upward into the developing eyelid to form a canaliculus and grows downward toward the nose to form the lacrimal sac and the nasolacrimal duct.

At approximately the third month of gestation, canalization of the nasolacrimal duct occurs because of a degeneration and shedding of the central cells of the duct, initially in the region of the lacrimal sac.

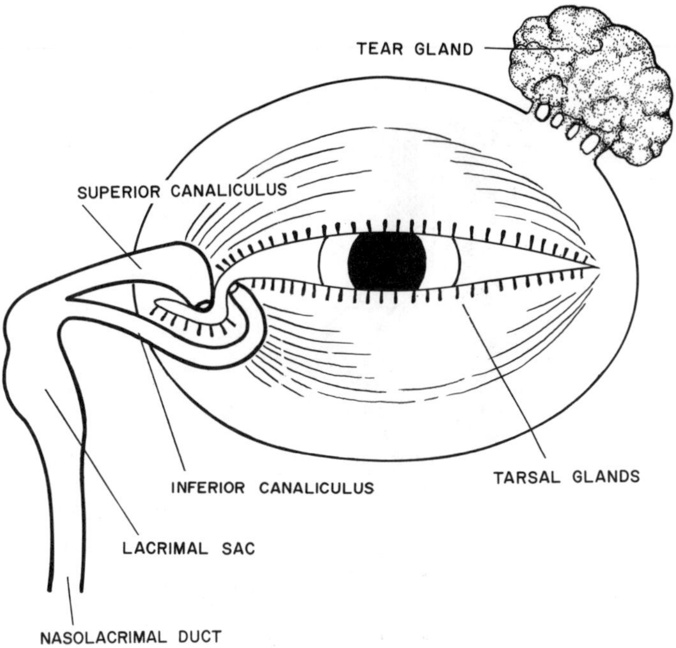

Fig. 8 Developing lacrimal apparatus, 7-month fetus.

Canalization of the nasolacrimal duct occurs at the punctum lacrimalia by the seventh month of gestation (Figs. 8, 9) and at the nasal cavity at birth or shortly thereafter, prior to tear formation. If spontaneous opening of the nasolacrimal duct by rupture of the nasolacrimal membrane does not occur by the time of tear formation (2 to 10 weeks postnatal), infection of the lacrimal sac (dacryocystitis) with epiphora and purulent discharge will ensue.

Causes of Dacryostenosis

What are some factors that may contribute to blockage of the nasolacrimal duct? Such blockage can result from: (1) a delayed separation of the nasolacrimal cord from the surface ectoderm and necrosis of the epithelial cord cells; (2) a retention of the nasolacrimal membrane between the nasolacrimal duct and the inferior meatus of the nasal cavity; (3) an obstruction of the nasolacrimal duct by folds of mucous membrane; (4) a partial occlusion of the duct by the turbinate bones; and (5) faulty cartilage development. In addition, the nasal ostium of the duct may be excessively small and incapable of normal function.

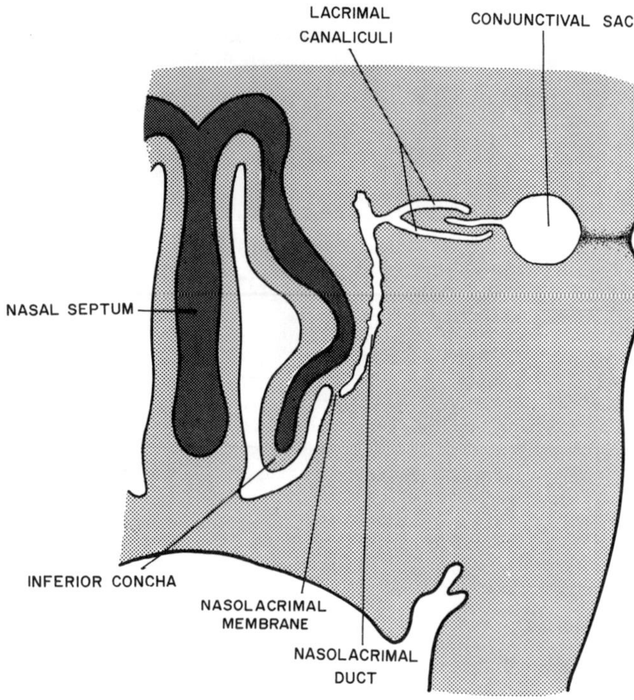

Fig. 9 Frontal section of a 7-month fetus showing the course of the nasolacrimal duct.

Associated Malformations

What malformations most commonly occur in conjunction with dacryostenosis? Congenital glaucoma, corneal foreign body, conjunctivitis, keratitis, and iritis most frequently accompany dacryostenosis. Impatency of the nasolacrimal duct also occurs in cleft lip and palate.

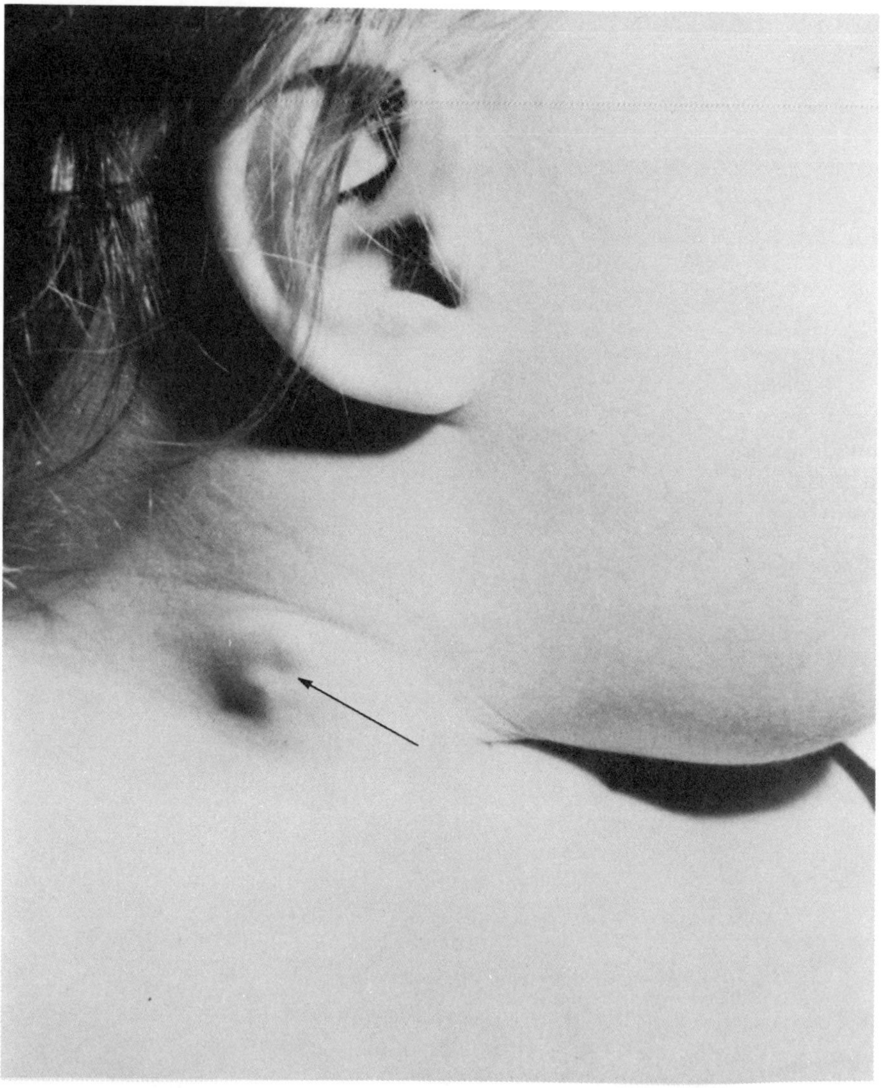

Right-sided branchial fistula of the second branchial cleft, *arrow*.

7

Branchial Fistula

The mother of a 15-month-old infant complained of a constant wetness on the baby's neck predominantly on the right side. Examination revealed a tiny opening on the right side on the lower third of the neck on the anterior border of the sternocleidomastoid muscle. No inflammation or swelling was present, although small amounts of mucoid material discharged from the opening each day. No additional malformations were noted. A family history revealed no known malformations of this type.

DIAGNOSIS

Right-sided branchial fistula (lateral cervical fistula, congenital sinus of the neck, persistent branchial cleft) of the second branchial (pharyngeal) cleft.

TREATMENT

Because of the constant wetness and the possibility of infection, surgical removal of the fistula was indicated. The infant was placed in a prone position with the shoulders elevated and the chin turned opposite to the fistula. A ureteric catheter was passed the entire length of the fistula and an incision was made around the external fistular opening along the cleft lines of the skin (Langer's lines). Blunt dissection remained close to the fistular tract to prevent injury to the vessels. The tract extended to the tonsillar fossa, and a second incision, parallel to the first and beneath the angle of the mandible, was made and the tract removed through the second incision. The skin was closed in layers and recovery was uneventful with no recurrence by 2 years of age.

DISCUSSION

Anomalies of the second branchial cleft are approximately 60 per cent more common than those of the first cleft.[27] In 1827 von Baer[20] described the branchial arches in man, and five years later Ascherson[1] related branchial fistulas to malformations of the branchial arches, pouches, and clefts. The term "branchial fistula" was introduced by Heusinger in 1864.[15] Wenglowski (1912)[49] proposed that not all fistulas or cysts were developmental in origin, while King (1949)[20] said that not all branchial cysts were branchial in origin.

Embryology

How do you explain embryologically the presence of a branchial fistula? Between the third and fourth weeks of gestation (Fig. 10A), the visceral mesoderm in the lateral wall and floor of the pharynx forms five or six pharyngeal arches. During development each arch acquires a cartilagenous skeleton, a muscular component, and a characteristic artery and nerve. Although the muscles of each arch migrate, they retain the nerve and arterial supply from their arch of origin. The pharyngeal arches are separated externally by four pharyngeal grooves lined with ectoderm; internally, corresponding to these grooves, are four pharyngeal pouches lined with entoderm. There is a fifth pharyngeal pouch, but it is atypical and usually considered as part of the fourth. The ectoderm and entoderm, with some interposed mesoderm, come into contact deep in the pharyngeal groove, forming a cleft (closing) membrane which normally does not open to the outside. Shortly, the entoderm withdraws from the ectoderm, and the pharyngeal pouches elongate to form dorsal and ventral wings that give rise to glandular structures.

At five-weeks' gestation (Fig. 10B), the second pharyngeal arch grows caudally over the third and fourth arches, partly enclosing them in an ectodermal pocket, the cervical or precervical sinus of His. Caudal to the arches is the epipericardial ridge in which the sternocleidomastoid and infrahyoid muscles develop. The cervical sinus is obliterated at the end of the second month by fusion of the second arch with the epipericardial ridge.

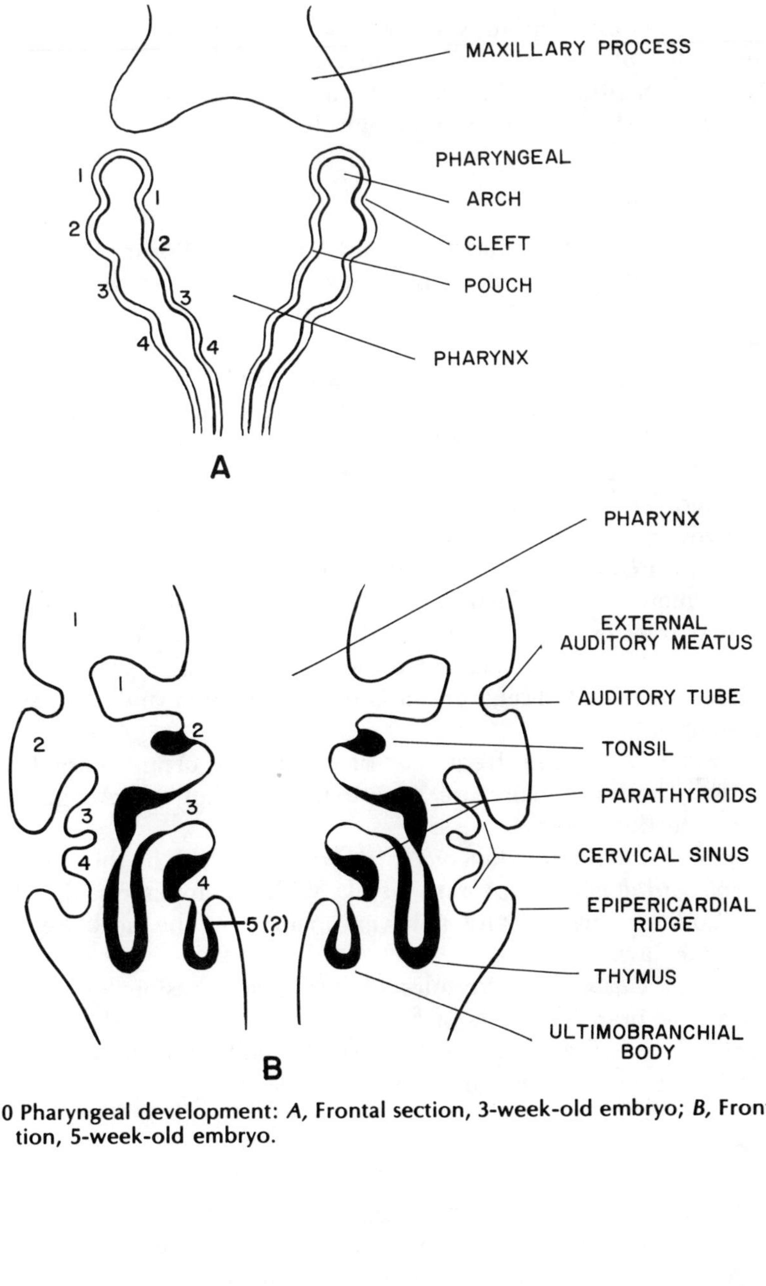

Fig. 10 Pharyngeal development: *A*, Frontal section, 3-week-old embryo; *B*, Frontal section, 5-week-old embryo.

If the cleft membranes should rupture (most frequently the second) before the entoderm withdraws from the ectoderm, there is a communication of the pharynx to the outside via the cervical sinus. This communication constitutes a branchial fistula and is usually immediately anterior to the sternocleidomastoid muscle.

Branchial Vestiges

What are the more frequent branchial malformations? For convenience each malformation is described separately; however, they are closely related and often arise simultaneously from the same branchial vestiges.

Branchial Cyst. Branchial cysts usually are unilateral, seldom have a familial incidence, and occur at about the level of the second (hyoid) arch deep to the anterior border of the sternocleidomastoid muscle on the carotid sheath. Many branchial cysts are smooth and completely encapsulated having no special attachments, although some possess an upper or lower pedicle.

External Fistula. Unlike cysts, fistulas are usually bilateral, the right side more often affected, and familial incidence is not unusual. The fistular tract opens in the lower third of the neck along the anterior border of the sternocleidomastoid muscle; the fistula is present at birth and exudes a mucoid discharge. Secondary infection and inflammation are frequent.

Internal Fistula. The frequency of occurrence of an internal fistula is unknown, because many are symptomless, merely discharging their secretion into the throat.

Complete Pharyngocutaneous Fistula. Described in this case.

Cervicoaural Fistula. A rare fistula with an upper opening in the external auditory meatus and a lower opening in the neck below the angle of the jaw.

Branchial Cartilages. Nodules of cartilage occasionally are found in the walls of branchial cysts or fistulas.

Cervical Auricles. Tags of skin occasionally protrude at the cutaneous orifice of a branchial sinus or fistula.

Classification

How are branchial malformations classified? Sutton (1887)[42] described four distinct types of branchial malformations: (1) complete fistula—external and internal openings; (2) incomplete external fistula —only external opening present; (3) incomplete internal fistula—only internal opening present; and (4) cyst—a closed canal externally and internally possessing an unobliterated canal segment in the neck. Bailey (1933)[2] classified lateral cervical cysts into four types as determined by the surrounding structures.

Associated Malformations

Some branchial malformations, especially fistulas, are associated with facial or cervical anomalies, e.g., preauricular sinuses or accessory auricular tubercles. Associated malformations usually occur on the same side as the branchial anomaly.

Table 2. Human pharyngeal region derivatives

	1	2	3	4	5	6
Skeletal Arch	Maxillary process Sphenomandibular lig. Malleus Incus Meckel's cartilage	Stapes, styloid process Stylomandibular and stylohyoid ligs. Lesser cornu hyoid Upper part of body of hyoid bone	Greater cornu hyoid Lower part of body of hyoid bone	Thyroid cartilage Thyrohyoid lig. (?) Cuneiform cart.	Not formed	Thyroid cart. (?) Corniculate cart. Arytenoid cart. Circoid cart.
Ectodermal Cleft	Ext. auditory meatus Outer layer tympanic membrane	Obliterated	Obliterated	Obliterated	Not formed	Not formed
Entodermal pouch	Tympanic cavity Inner layer tympanic membrane Lining mastoid air cells Auditory tube	Dorsal wing—tympanic recess Ventral wing—tonsilar fossa, palatine tonsil	Dorsal wing—inferior parathyroid Ventral wing—thimus	Dorsal wing—superior parathyroid Ventral wing—thymus, lateral thyroid (?)	Ultimo-branchial body (?)	Not formed
Arch Musculature	Mus. of mastication Ant. belly digastric Tensor velli palatini Tensor tympani Mylohyoid	Mus. of facial expression Post. belly digastric Stylohyoid Stapedius	Pharyngeal con-strictors—up-per, middle (?) Stylopharyngeus	Pharyngeal constrictors (?) Cricothyroid Muscles of larynx (?)	Not formed	Intrinsic of larynx Inf. pharyngeal constrictor Upper fibers of esophagus
Arch Arterial	Maxillary	Stapedial	Stem L. internal carotid L. common carotid	Arch of aorta Right subclavian	Not formed	Pulmonary Ductus arteriosus
Arch Nerve	V_3—(Mandibular of Trigeminal)	VII—(Facial)	IX—(Glosso-pharyn-geal)	X—(Superior laryngeal of vagus)	Not formed	X—(Recurrent laryngeal of vagus)

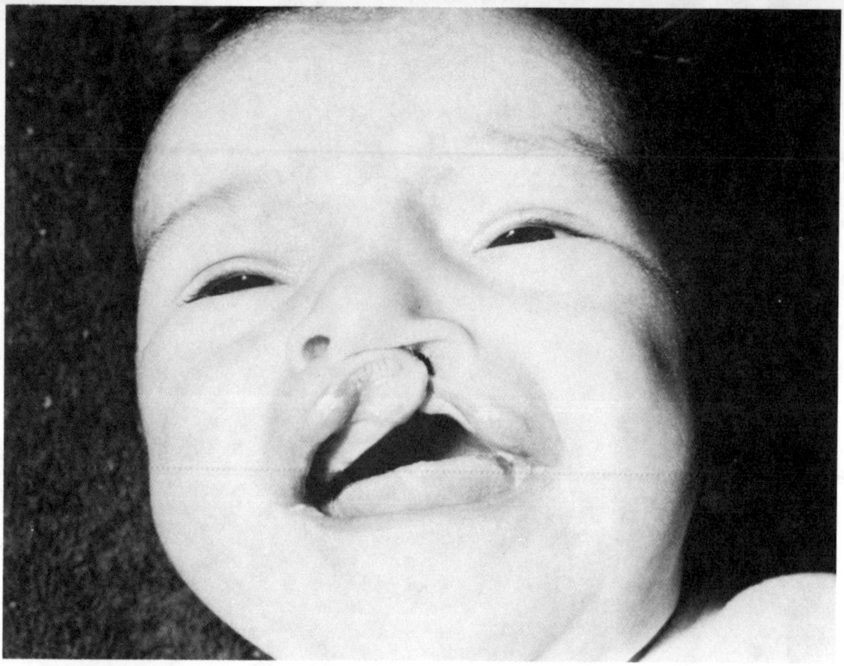

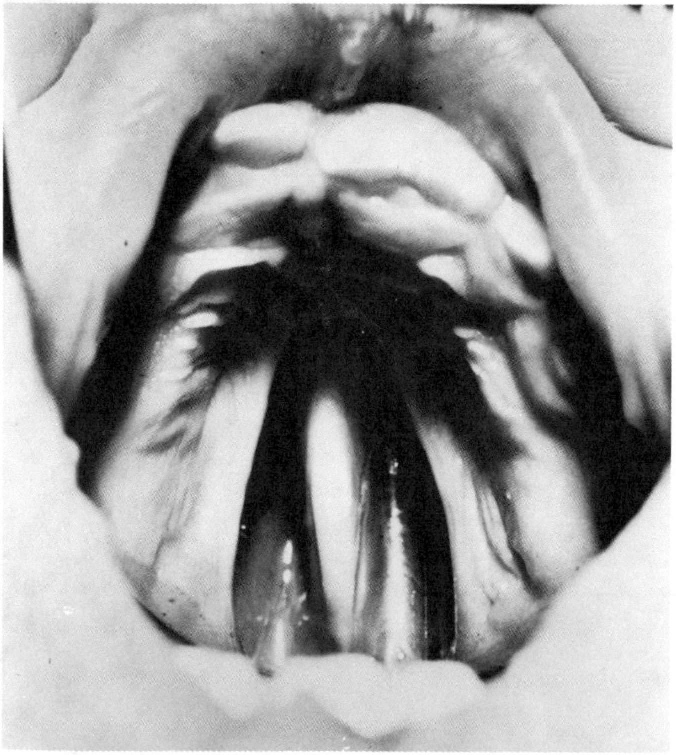

Above, Left upper unilateral cleft lip; *Below,* Complete midline cleft of the palate.

8

Cleft Lip and Palate

Initial examination of a full-term Caucasian male infant revealed a cleft lip. Further exploration disclosed involvement of the alveolar process and palate. No other malformations were noted. Feeding was not an immediate problem and surgical repair was delayed. Family history revealed a bifid uvula in a female sibling.

DIAGNOSIS

Left upper unilateral cleft lip and complete midline cleft of the palate.

TREATMENT

When the infant was six weeks of age, repair of the lip, alveolus, and premaxilla was performed. The palate was repaired 14 months after birth, prior to the development of numerous speech habits. Additional surgery of thc lip or nose, as a result of growth, were to be made when and if indicated. Ear, nose, and throat infections were carefully watched. Follow-up continued to the fifteenth year to insure articulate speech, adequate and functional dental care, and healthy social adjustment.

DISCUSSION

Progress in cleft lip repair is divided into three periods. The first period began with Paré (1634),[18] the first to describe in detail the principles and techniques of surgical treatment in cleft lip, and ended with Velpeau (1838).[47] Period two, which began with Mirault (1844)[26] and ended with Blair and Brown (1930),[4] witnessed the inception of anesthesia, improved antisepsis, the technique of interrupted sutures and improved suture material, and the introduction of modern surgical instruments. The third period continues to the present and is characterized by additional new methods of repair (LeMesurier, 1949;[21] Millard, 1958;[25] Steffenson, 1949;[40] and Tennison, 1952[43]).

Repair of cleft palate was begun during the Renaissance by Franco (1561),[13] who described the congenital origin of this malformation. In the nineteenth century,[6, 9] rapid progress was made in successful closure of the cleft, and shortly before World War I, surgery, aided by roentgenograph and growth-pattern studies, delineated basic techniques. Great strides have been made in management of cleft palate since World War II, and today care of the cleft palate patient involves a group of coordinated specialists.

Statistics on the occurrence of facial clefts vary with the country and the racial group. Among Caucasians, cleft lip with or without cleft palate occurs in 0.94 per 1,000 live births, with 51 to 59 per cent of the clefts on the left side. Approximately two thirds of those affected have both cleft lip and cleft palate.[36]

Embryology

Discuss development of the facial and palatal regions. The facial region develops between the fourth and eighth week of gestation and is dominated by changes resulting in the formation of the nose. Palatal development occurs between the seventh and twelfth week of gestation and is divided into formation of the primary palate, which includes the prolabium, premaxilla, and cartilaginous septum, and formation of the secondary palate, which consists of the hard and soft palates. The primary and secondary palates develop as individual entities and are separated by the incisive foramen.

By the fourth week of gestation (Fig. 11A), five processes (mesenchymal thickenings), separated by grooves, surround the stomodeum: two maxillary processes, two mandibular processes, and a single frontonasal process.

During the fifth week (Fig. 11B), olfactory placodes (epithelial thickenings) develop lateral to the frontonasal process and cranial to the stomodeum. The placodes gradually deepen and form olfactory pits as a result of the elevation of the surrounding ectoderm caused by elevations of the underlying mesoderm. On either side of the pit, the proliferations are called the medial and lateral nasal processes. The me-

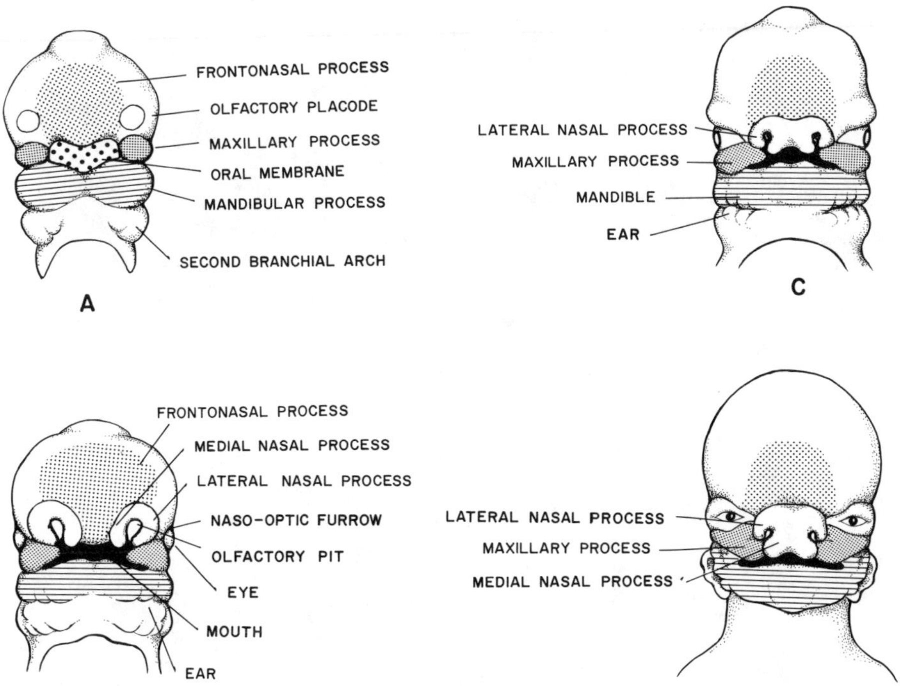

Fig. 11 Developing primary palate and face: *A,* 4-week-old embryo; *B,* 5-week-old embryo; *C,* 6-week-old embryo; *D,* 8-week-old embryo.

dial nasal process becomes incorporated into the frontonasal process, after which the mesenchyme of the frontonasal process proliferates and migrates over the stomodeal margin to where the stomodeal membrane ruptured; this forms the primary palate (prolabium, premaxilla).

In the sixth week (Fig. 11C), the maxillary and lateral nasal processes merge as a result of proliferation and penetration of the underlying mesoderm, thus obliterating the naso-optic furrow. Concurrently, the two mandibular processes fuse.

Between the seventh and eighth week (Fig. 11D), the upper lip is completed by maxillary mesoderm which merges at the midline. The maxillary mesoderm that forms the upper lip may bury the primary palate at the stomodeal margin, after which the primary palate no longer contributes to the upper lip but to the alveolus. Continued merging lateral of the maxillary and mandibular processes forms the cheeks.

Near the completion of the primary palate in the seventh week

(Fig. 12A, C), two vertical palatal processes from the maxilla extend from the primary palate to the tonsillar pillars.

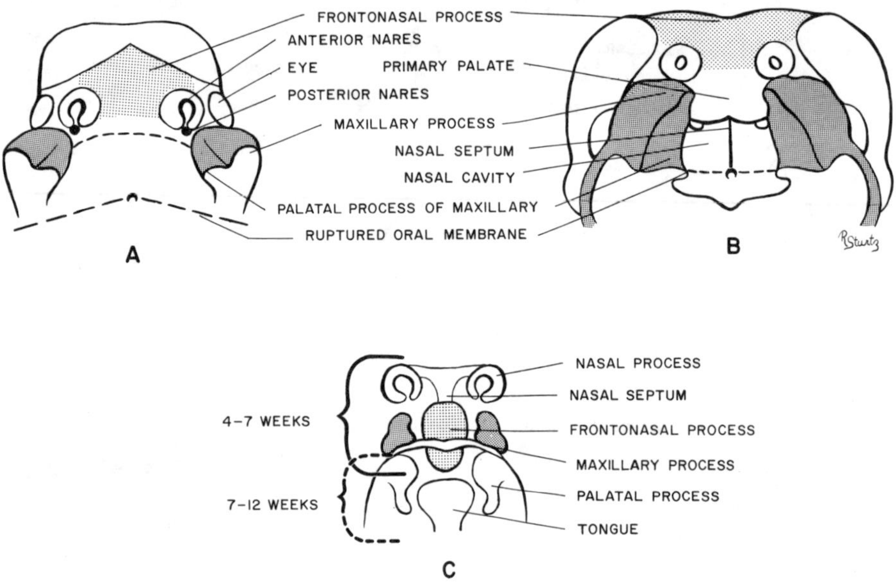

Fig. 12 Roof of the stomodeum showing secondary palate development: *A*. 7-week-old embryo. *B*. 8- to 9-week-old embryo. *C*. Composite drawing of roof of stomodeum showing secondary palate development at 4 to 7 weeks and at 7 to 12 weeks gestation.

Between the eighth and ninth week (Fig. 12B, C), these palatal processes attain a horizontal position and for a short time are separated by the tongue.

Due to the formation of the neck in the tenth to twelfth week, the tongue comes to lie in the oral cavity, allowing the palatal processes to fuse, thus forming the secondary palate.

Fusion Processes

What mechanisms are involved in palatal (primary and secondary) development? The classic studies of Dursy (1869)[10] and of His (1874)[16] describe the normal development of the palate as occurring from a series of mesodermal processes; these processes are covered by ectoderm and separated by open spaces or clefts. The processes fuse as a result of abutment of the surface ectoderm, which subsequently breaks down at the points of fusion, leaving a solid mesodermal union. At present this theory is considered inadequate, and Stark (1954)[39] describes facial development as occurring from a number of mesodermal masses which underlie the ectoderm so as to form ectodermal furrows. Merging of the mesodermal masses smooths the ectodermal furrows and results in a continuous structure of mesoderm. This concept is applicable to facial, lip, and primary palate development, but not to the secondary palate development, which forms by ectodermal fusion.

Insufficient mesoderm results in structural or functional clefts of the palate. Clefts of the secondary palate may result from: (1) insufficient mesoderm to elevate the palatal shelves prior to their union and before the pre-existing cleft is widened by transverse growth of the skull; (2) inability of the palatal shelves to bridge an abnormally wide skull; and (3) interposition of the tongue in an overcrowded oral cavity due to retarded growth of the neck.

Classification

How have cleft lip and cleft palate been classified? Recognizing three major groups of deformities, Davis and Ritchie (1922)[7] proposed the following:

Group I —prealveolar clefts
Group II —postalveolar clefts
Group III—1) alveolus plus cleft lip
—2) alveolus cleft with a normal palate

Scientific classification by Veau in 1931[46] indicated the degree of deformity by a numerical scale of I to IV; the greater the number, the more deviation from the norm.

The classification of Kernahan and Stark (1958)[19] (Fig. 13) relates the congenital deformity to embryological development; in 1967 it was adopted by the International Confederation for Plastic Surgery.

A. Clefts lying anterior to incisive foramen—involves primary palate.
B. Clefts lying posterior to incisive foramen—involves secondary palate.
C. Clefts formed by failure of normal development of both primary and secondary palates.

Following Kernahan and Stark's report, the Nomenclature Committee of the Cleft Palate Association proposed an additional fourth group to include facial clefts.

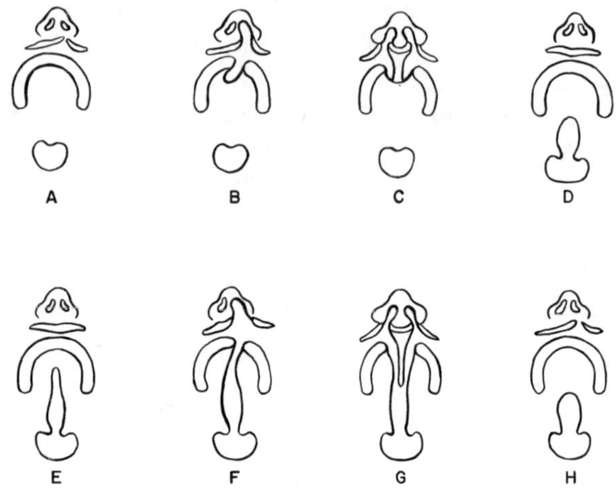

Fig. 13. Cleft palate classification: *A,* Left unilateral incomplete cleft of the primary palate; *B,* Left complete cleft of the primary palate ending at the incisive foramen; *C,* Bilateral complete cleft of the primary palate; *D,* Incomplete cleft of the secondary palate (when alone); *E,* Complete cleft of the secondary palate (soft and hard palates); *F,* Left complete cleft of the primary and secondary palates; *G,* Bilateral complete cleft of the primary and secondary palates; *H,* Left incomplete cleft of the primary palate and incomplete cleft of the secondary palate. (From Kernahan, D. A., and Stark, P. B.: Plast. Reconstr. Surg. 22:435, 1958, with permission.)

Treatment Objectives

What are the major objectives in repair of cleft lip? Desired results include: (1) a symmetrical lip with a natural-appearing cupid's bow; (2) a natural-appearing philtral ridge and dimple; (3) negligible scarring; (4) a symmetrical nose with restoration of the nostril floor and elongation of the columella; and (5) a full vermilion with preservation of the mucocutaneous line.

What is the major objective in repair of cleft palate? Satisfactory repair constructs a competent, functioning, air- and watertight valve at the junction of the soft palate and pharynx; repair is performed early enough to allow the child to begin speech with a functioning velopharyngeal valve, free of edema. Additional objectives are: (1) preservation of normal hearing; (2) normal development and bone growth of the central face; and (3) functional and attractive dentition.

Associated Malformations

What malformations are closely associated with clefts of the lip and palate? There are two groups of congenital defects correlated with cleft lip and palate; (1) those occurring in the immediate vicinity of the defect, i.e., Pierre-Robin syndrome, Klippel-Feil syndrome, and Psaume syndrome; and (2) those occurring elsewhere in the body, i.e., congenital heart, clubfoot, syndactyly, and polydactyly.

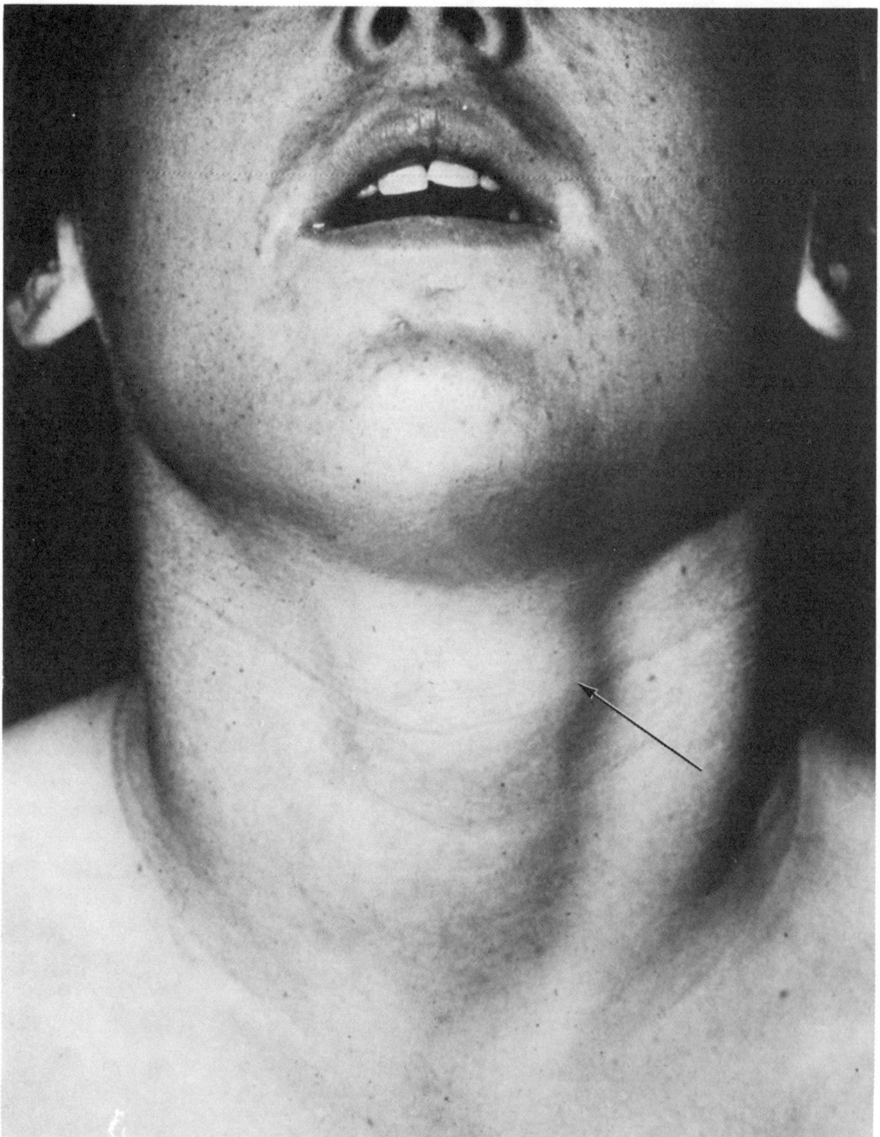

Infrahyoid thyroglossal duct cyst, *arrow*.

9

Thyroglossal Duct Cyst

A 4-year-old girl is brought to a physician's office with an acutely inflamed, soft, midline, infrahyoid swelling. The mother states that the swelling was first noticed in conjunction with the inflammation. The swelling shows no apparent secretion and, upon oral examination with the tongue protruded, is observed to move slightly craniad. Surgery is performed once the local infection, which is treated with antibiotics, has subsided.

DIAGNOSIS

Infrahyoid thyroglossal duct cyst with tract.

TREATMENT

With the head hyperextended, a nasotracheal anesthesia was given to allow for palpation of the thyroglossal duct termination in the floor of the mouth. A transverse incision (along Langer's lines) over the 2-cm. cyst allowed it to be easily identified. From the cranial end of the cyst a grayish well-defined thyroglossal duct was traced to where it entered the midportion of the hyoid bone. Approximately 1 cm. of the middle of the bone was removed to insure extirpation of the duct; from the hyoid bone to the foramen cecum, the duct became friable and poorly identified, and to insure its complete elimination, a core of muscular tissue was excised, thus exposing the thyrohyoid membrane. The foramen cecum was then palpated to determine whether the entire duct had been removed. The thyroid gland was in its normal position, anterior to and on either side of the trachea; no accessory thyroid ma-

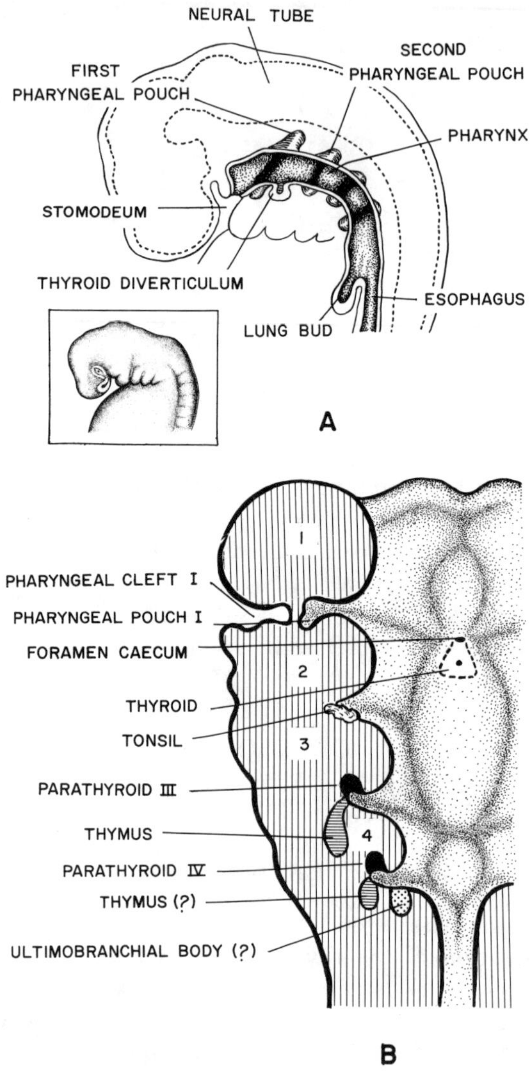

Fig. 14 Pharyngeal derivatives: *A,* Sagittal section, 4-week-old embryo; *Inset:* External view: *B,* Frontal section, 4-week-old embryo.

terial was found. The neck was closed in layers and recovery was uneventful with no recurrence of the cyst by the seventh year.

DISCUSSION

His (1874)[16] was the first to describe the origin of the human thyroid gland as developing from three components: (1) a median anlage from the pharyngeal floor, and (2) a pair of lateral pharyngeal components. Subsequent investigators[11, 29] have been interested in the abnormal locations and pathology of the thyroid.

Embryology

Discuss the normal developmental sequence of the thyroid gland. The thyroid is the first pharyngeal derivative to appear. This occurs toward the end of the fourth week of gestation (Fig. 14A, B) when the entoderm on the floor of the developing pharynx between the first and second pharyngeal pouches begins to proliferate and forms a median diverticulum. This diverticulum elongates and grows anteriorly and caudally with its most distal portion developing into a hollow vesicle, the thyroid primordium. The thyroid primordium grows in a lateral direction, but remains attached to the pharynx at the foramen cecum by the thyroglossal duct.

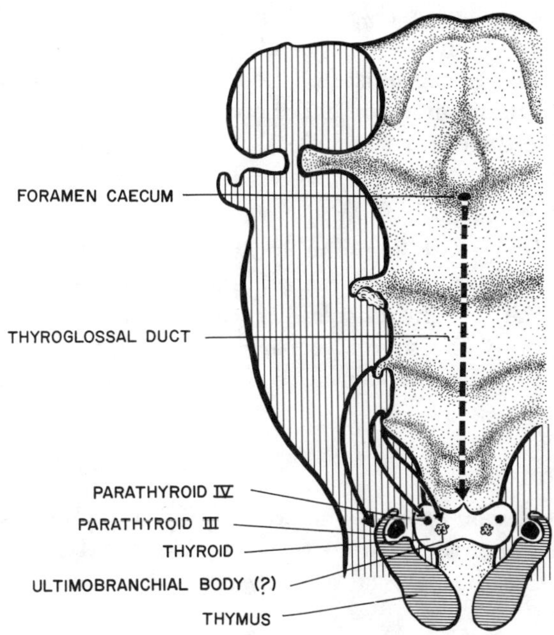

Fig. 15 Pharyngeal derivatives. Frontal section, 7-week-old embryo.

In the seventh week (Fig. 15), the thyroid incorporates the caudal diverticulae of the fourth pharyngeal pouches. These caudal diverticulae have been named rudimentary fifth pouches, ultimobranchial bodies, or lateral thyroid primordia.[31] The thyroid continues to grow and by the end of the seventh week reaches its final location anterior to the first six tracheal cartilages.

Subsequently, the second branchial arches meet anteriorly in the midline to form the hyoid bone. In the adult, the thyroglossal duct is either posterior to or passes through the hyoid bone.

In the eighth week, the thyroglossal duct is obliterated and the thyroid loses its attachment to the pharynx. A part of the caudal end of the thyroglossal duct may persist to form a pyramidal lobe; this occurs in one third to one fourth of normal individuals.[27]

During the tenth week, thyroid follicles form, but colloid is not visible until the end of the third month when the gland becomes functional.

Etiology

Explain the etiology of a thyroglossal duct cyst. The cyst is a remnant of the thyroglossal duct and may be found anywhere along the path of the growing thyroid from the base of the tongue to the suprasternal notch. The cyst may enlarge, become infected, and/or develop a connection to the outside called a thyroglossal fistula. The cysts are discovered in infancy, in childhood, or occasionally at birth, and may recur if the duct is not completely excised.

Cyst Location

What are the most common locations of a thyroglossal cyst? A cyst most often arises: (1) in the midline below the hyoid bone in association with the thyrohyoid membrane; (2) on either side of the midline in the region of the thyroid cartilage; (3) on the cricothyroid membrane; and (4) above the thyroid bone. Rarely a cyst develops at the base of the tongue.

Diagnostic Errors

What errors can be made in identifying a thyroglossal duct cyst? Dermoid cysts, lipomas, ectopic thyroid tissue, and lymph nodes have been identified as thyroglossal duct cysts.

Associated Malformations

Additional malformations are seldom found with a thyroglossal duct cyst, although the epithelium of the duct can develop most diseases occurring in the gland itself (adenoma, carcinoma, thyroiditis, hyperthyroidism, and recurrent hyperthyroidism).[27]

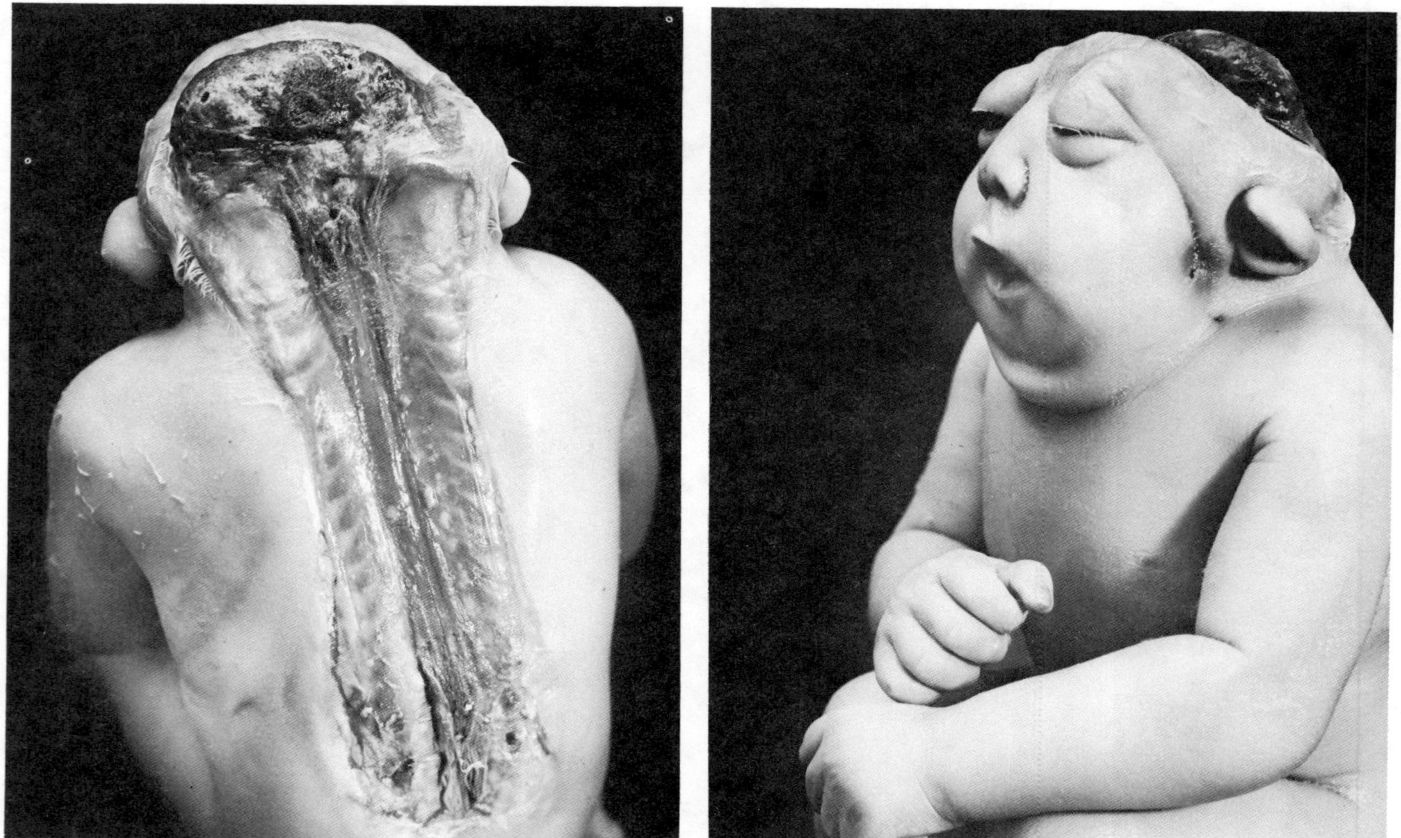

Left, Anencephalic. (From Sucheston, M.E., and Cannon, M S.: J. Obstet. Gynecol. 35:544, 1970, with permission.) *Right,* Craniorachischisis.

10

Anencephaly

In the thirty-seventh week of gestation, a 32-year-old gravida IV, para I mother was admitted to the hospital with hydramnios. She was type O, Rh positive with a family history of frequent abortions. Transabdominal amniocentesis showed results indicative of anencephaly and esophageal atresia. Anencephaly was diagnosed by x-ray. Labor was induced, and an anencephalic female weighing 2,100 gm. was delivered, but died 10 minutes after birth. The gross features of the newborn were: (1) absence of a cranial vault; (2) spinal canal completely open; (3) forebrain and midbrain replaced by a spongy mass of tissue, the area cerebrovasculosa; (4) thick, low-set, and protruding ears; (5) absence of a neck with the chin over the sternum and a reduced thoracic cavity; (6) bulging eyes due to a foreshortening of the anterior cranial fossa; and (7) a protruding tongue, all of which are characteristic of anencephaly. Additional malformations were a cleft palate and a cervical fistula.

DIAGNOSIS

Dysraphia—Anencephaly with craniorachischisis. Condition incompatible with life.

DISCUSSION

Lycosthenem[22] first described a female anencephalic in France in 1557, while von Recklinghausen (1886)[32] was the first to maintain that anencephaly was a primary arrest of closure of the neural tube.

Four major areas of interest regarding anencephaly have developed: (1) gross embryologic and fetal development of the monster;[14, 23, 30] (2) associated malformations;[17, 28, 48] (3) etiologic and genetic prevalence of anencephaly in successive pregnancies;[8, 12, 44] and (4) pituitary, adrenal, and gonadal development and interrelationship.[3, 41, 50] Anencephaly occurs in 1 out of 1,000 births.[36]

Embryology

Discuss the early formation of the central nervous system and its skeletal covering. The neural plate appears at three-weeks' gestation (Fig. 16A) as a thickened ectodermal area, which becomes distinguish-

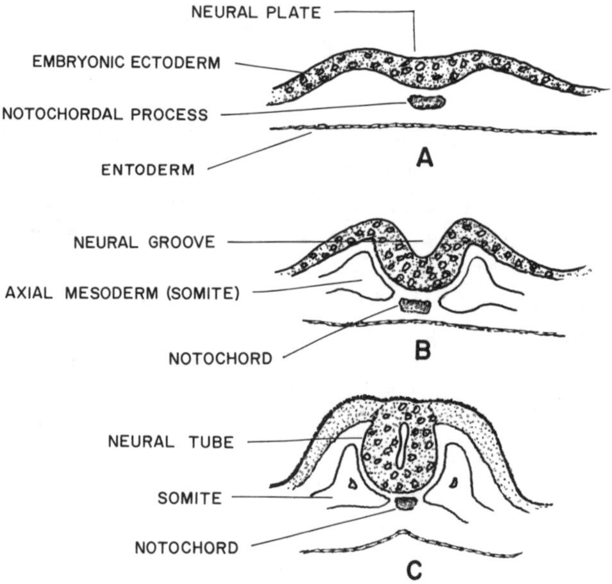

Fig. 16 Neural tube formation: *A* and *B*, 3-week-old embryo; *C*, 4-week-old embryo.

able by an increase in the height of its cells as a result of an inductive influence from underlying mesodermal structures. The plate overlies the notochordal process and the medial edge of the paraxial mesoderm and becomes depressed axially to form a shallow neural groove (Fig. 16B). In early somite embryos, in the future cervical region, the lateral margins of the neural groove meet and fuse to form the neural tube. The primordium of the future brain is immediately cranial to the cervical region, while the spinal cord is caudal. By the fourth week (Fig. 16C), the neural tube has closed except at the cranial (anterior) and caudal (posterior) neuropores; closure here occurs at 24 and 26 days, respectively. By the end of the somite stage, the central nervous system is represented by a closed tube, dilated cranially and bent in conformity with the dorsal convexity of the embryo.

The skeletal covering of the central nervous system consists of the skull and vertebral column. The skull is composed of the neurocranium surrounding the brain and the viscerocranium of the face and pharynx. The neurocranium appears at four-weeks' gestation as a mesodermal membrane that surrounds the already formed neural tube. The neurocranium is further divided into: (1) a cartilaginous portion—the primordia of the base of the skull and the sensory capsules, and (2) a membranous portion—the primordia of the skull cap, i.e., frontal, parietal, part of occipital, and squamotemporal bones. The vertebral column arises in the fourth week of gestation (Fig. 17A) from sclerotomic mesen-

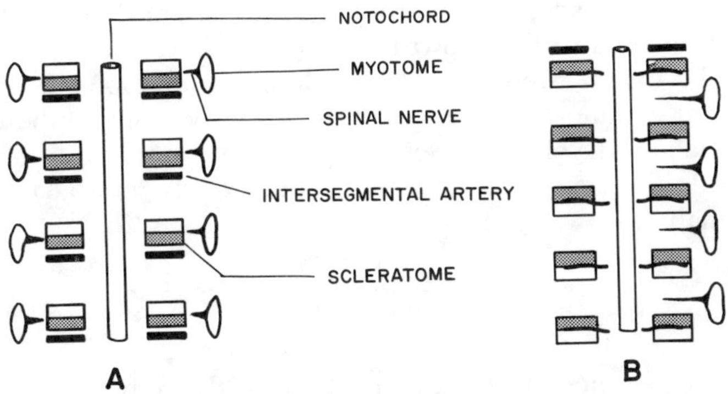

Fig. 17 Vertebral formation: *A,* 4-week-old embryo; *B,* 6-week-old embryo.

chyme derived from the axial mesoderm. The sclerotomes lie in pairs on either side of the notochord, each pair separated by an intersegmental artery. By the sixth week (Fig. 17B) each sclerotome is divided into a cranial and a more dense caudal portion. The caudal portion of a single sclerotome recombines with the cranial portion of the sclerotome just caudal to it. These recombined sclerotomes are the primordia of the definitive vertebrae. The recombined sclerotomes then grow in three directions: (1) mesad—to surround the notochord and form the vertebral body; (2) dorsad—to flank the neural tube and form the vertebral arch; and (3) vertrolaterad—to form the costal processes in the interspaces between divisions of myotomes. Mesenchymal tissue from the ends of two abutting vertebrae condenses to form the intervertebral disk; remnants of the notochord are incorporated into the disk and persist as the nucleus pulposus. As a result of vertebrae and intervertebral disk formation: (1) the original intersegmental arteries pass over the body of the vertebrae; (2) the segmental nerves from the myotomes lie at the level of the intervertebral disks; and (3) the myotomes and definitive vertebrae alternate in position.

Causes

Discuss embryologically the possible causes of anencephaly. Anencephaly, complete or almost complete absence of forebrain and midbrain, absence of skull, overlying membranes, and skin, and varying degrees of rachischisis, may result from: (1) primary arrest in closure of the neural tube at the cranial neuropore; (2) secondary opening and destruction of the closed neural tube; (3) abnormal flexures of the brain due to differences in the growth rates of the internal and external portions of the neural tube; (4) total failure of development of the skull or vertebral column as a consequence of faulty neural tube formation; (5) anomalous development of the cerebrovascular tree; (6) general ectodermal damage; (7) healing neural clefts or; (8) embryonic microcephaly.

Dysraphia

What experimental devices can produce dysraphia? Certain strains of mice are consistently born with dysraphia. X-irradiation produces chromosomal translocation that can result in dysraphia. Fasting or vitamin deficiency, anoxia, chemicals (i.e., trypan blue, salicylate, actinomycin, or hormones), or surgical opening of the newly closed neural tube can result in dysraphia.

Intrauterine Diagnosis

If the fetal head is not felt upon vaginal examination, there are a number of explanations: (1) the patient is obese; (2) an anterior placenta covers the head; (3) the fetal head has descended into the pelvis; or (4) the fetus is anencephalic or microcephalic.

Associated Malformations

What malformations occur in conjunction with anencephaly? Amyelia, facial or palatal clefts, visceral anomalies, and abdominal or diaphragmatic malformations are the more common defects accompanying anencephaly.

REFERENCES

1. Ascherson, F. M.: De Fistulis Colli Congenitis. C. H. Jonas, Berlin, 1832.
2. Bailey, J.: The clinical aspects of branchial fistulae. Brit. J. Surg. 21:173, 1933.
3. Benirschke, K.: Adrenals in anencephaly and hydrocephaly. Obstet. Gynec. 8:412, 1956.
4. Blair, V. P., and Brown, J. B.: Mirault operation for single harelip. Surg. Gynec. Obstet. 51:81, 1930.
5. Born, G.: Über die Nasenhöhlen und den Thränennasengang der Amphibien. Morp. Jahrbuch, Bd. II, 1876.
6. Cusack, J., and Crampton, P.: A case of cleft palate successfully operated on by Dr. James Cusack, and two cases, on which the operation was successfully performed by Sir Philip Crampton. Dublin J. Med. Sci., p. 321, 1843.
7. Davis, J. S., and Ritchie, H. P.: Classification of congenital clefts of the lip and palate. J.A.M.A. 79:1323, 1922.
8. DeSA, D. J., Strong, S. J., and Corney, G.: Anencephaly in one of monozygous twins. Biol. Neonat. 7-8:328, 1964-1965.
9. Dieffenbach, J. F.: Ueber das Gaumensegel des Menschen und der Säugethiere. Litt. Ann. Ges. Heilk. 4:298, 1826.
10. Dursey, E.: Zur Entwicklungsgeschichte des Kopfes des Menschen und der Höhern Wirbelthiere. Verlag der H. Lauppschen Buchlandlung, Tübingen, 1869.
11. Erdheim, J.: I. Über Schilddrüsenaplasie. II. Geschwülste des Ductus Thyreoglossus. III. Über einige menschliche Kiemenderivate. Beitr. Path. Anat. 35:366, 1904.
12. Feuerlicht, N.: Recurrent anencephaly. J.A.M.A. 143:23, 1950.
13. Franco, P.: Bec de lièvre. In Nicaise, E. (ed.): Chirurgie Composée in 1561. Alcon, Paris, 1895.
14. Giroud, A.: Causes and morphogenesis of anencephaly. Ciba Foundation Symposium of Congenital Malformations. Little, Brown and Company, Boston, 1960.

15. Heusinger, R.: Hals-kiemen-fisteln von noch nicht beobachteter Form. Virchow Arch. Path. Anat. 29:358, 1864.
16. His, W.: Unsere Körperform und Das Physiologische Problem Ihrer Entstehung. Verlag von F. C. W. Vogel, Leipzig, 1874.
17. Hurwitz, C. H.: Anencephalus. Obstet. Gynec. 6:303, 1955.
18. Johnson, T.: The works of that famous Chirurgion Ambrose Parey. Translated out of Latine and compared with the French. Th. Cotes and R. Young, London, 1634.
19. Kernahan, D. A., and Stark, R. B.: A new classification for cleft lip and cleft palate. Plast. Reconstr. Surg. 22:435, 1958.
20. King, E. S. J.: The lateral lympho-epithelial cyst of the neck ("branchial" cyst). Aust. New Zeal. J. Surg. 19:109, 1949.
21. LeMesurier, A. B.: A method of cutting and suturing the lip in the treatment of complete unilateral clefts. Plast. Reconstr. Surg. 4:1, 1949.
22. Lycosthenem, C.: Prodigiorum ac Ostenorum Chronicon. Basileae per Henricum Petri. Mense Augusto, 1557.
23. Mall, F. P.: A study of the causes underlying the origin of human monsters. J. Morph. 19:3, 1908.
24. Meller, J.: Diseases of the lacrymal apparatus. Trans. Ophthal. Soc. UK 49:233, 1929.
25. Millard, D. R., Jr.: A radical rotation in single harelip. Am. J. Surg. 95:318, 1958.
26. Mirault, G.: Lettre sur l'operation du bec-de-lièvre. J. Chir. (Malgaigne) p. 237, Sept., 1844.
27. Mustard, W. T., Ravitch, M. M., Snyder, W. H., Jr., Welch, K. J., and Benson, C. D. (eds.): Pediatric Surgery. Year Book Medical Publishers, Inc., Chicago, 1969.
28. Nichols, J., and Schrepfer, R.: Polyhydramnios in anencephaly. J.A.M.A. 197:549, 1966.
29. Nicholson, G. W.: Studies on tumor formation. II. Tissue malformations. Anomalies of bulk and of differentiation. Guy Hosp. Rep. 72:193, 1922.
30. Norman, A. P.: Congenital Abnormalities in Infancy. Blackwell Scientific Publications, Oxford, 1963.
31. Patten, B. M.: Human Embryology, ed. 3. McGraw-Hill Book Company, New York, 1968.
32. von Recklinghausen, F.: II. Ueber die Art und die Entstehung der Spina bifida, ihre Breziehung zur Rückenmarks—und Darmspalte. Virchow Arch. Path. Anat. 105:296, 1886.

33. Reiling, P., and Bowen, S. F., Jr.: Congenital nasolacrymal duct obstruction in infancy. Missouri Med. 66:881, 1969.
34. Riser, R. O.: Dacryostenosis in children. Amer. J. Ophthal. 18:1116, 1935.
35. Rochon-Duvigneaud: Dilatation des voies lacrymales chez le foetus et le nouveau-né consécutive à l'imperforation de leur orifice inférieur. Conditions anatomiques qui favorisent la dacryocystite congénitale. Arch. Ophtalmol. (Paris) 19:81, 1899.
36. Rubin, A.: Handbook of Congenital Malformations. W. B. Saunders Company, Philadelphia, 1967.
37. Schaeffer, J. P.: The genesis and development of the nasolacrimal passages in man. Am. J. Anat. 13:1, 1912.
38. Stanculeanu, G.: Recherches sur le developpement des voies lacrymales chez l'homme et chez les animaux. Arch. Ophtalmal. (Paris) 20:141, 1900.
39. Stark, R. B.: The pathogenesis of harelip and cleft palate. Plast. Reconstr. Surg. 13:20, 1954.
40. Steffensen, W. H.: A method for repair of the unilateral cleft lip. Plast. Reconstr. Surg. 4:144, 1949.
41. Sucheston, M. E., and Cannon, M. S.: Microscopic comparison of the normal and anencephalic human adrenal gland with emphasis on the transient-zone. Obstet. Gynec. 35:544, 1970.
42. Sutton, J. B.: On branchial fistulae, cysts, diverticula, and supernumerary auricles. J. Anat. Physiol. 21:289, 1887.
43. Tennison, C. W.: The repair of the unilateral cleft lip by the stencil method. Plast. Reconstr. Surg. 9:115, 1952.
44. Thoms, H.: Family prevalence in anencephaly. J.A.M.A. 70:10, 1918.
45. Vannas, S., and Liesmaa, M.: Incidence of congenital anomalies of the lacrimal passages. Acta. Ophthal. 42:962, 1964.
46. Veau, V.: Division Palatine. Masson, Paris, 1931.
47. Velpeau, A. L. M.: New Elements of Operative Surgery, Vol. 3, translated by P. S. Townsend. Samuel S. and William Wood, New York, 1847.
48. Vogel, F. S.: The anatomic character of the vascular anomalies associated with anencephaly. Am. J. Path. 39:163, 1961.
49. Wenglowski, R.: Über die Halfisteln und cysten. Arch. Klin. Chir. (Berlin) 98:151, 1912; 100:789, 1912-1913.
50. Zondek, L. H., and Zondek, T.: The secretory activity of the human epididymis in anencephaly. Ann. Paediat. (Basel) 204:301, 1965.

THORAX

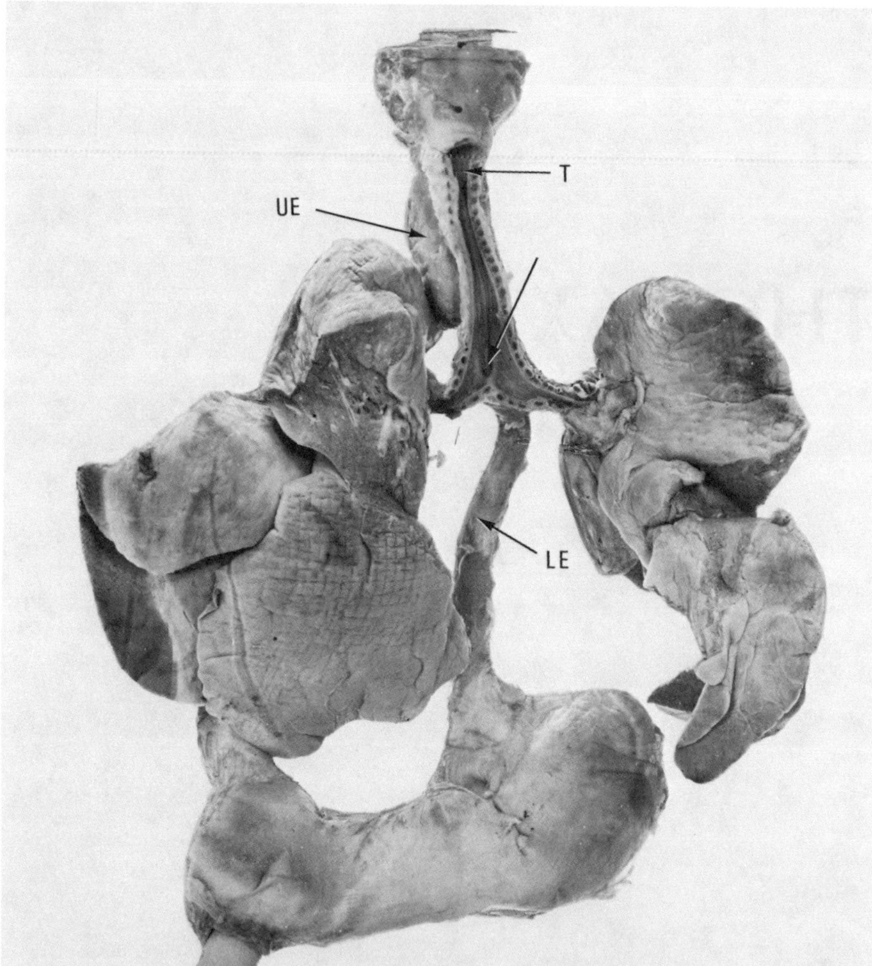

Tracheoesophageal fistula, *arrow*. Trachea, *T;* upper esophageal pouch, *UE:* lower esophageal pouch, *LE*.

11

Congenital Esophageal Atresia with Tracheoesophageal Fistula

A 4-lb., 5-oz. premature female infant was born of a pregnancy complicated by maternal hydramnios. At birth and shortly thereafter, the infant showed an excessive amount of oropharyngeal secretion. Initial feeding twelve hours after birth resulted in vomiting, choking, and cyanosis. The abdomen was distended and became larger upon coughing and crying. The infant was subsequently transferred to an intensive care unit.

EXAMINATION

The possibility of esophageal atresia accompanied by tracheoesophageal fistula resulted in the introduction of a radiopaque catheter into the esophagus. The catheter stopped 9 cm. from the upper gum margin. A roentgenogram was made to determine the level of the obstruction in relation to the vertebrae and to enable estimation of the length of the blind upper esophageal pouch. Due to the distended abdomen, a full-length air contrast roentgenogram was made, which indicated air in the stomach and intestines, thus establishing the presence of a fistulous communication between the trachea and the lower esophagus. The roentgenogram of the abdomen revealed no additional malformations and the infant was considered a good risk for a one-stage surgical repair.

DIAGNOSIS

Congenital esophageal atresia with tracheoesophageal fistula (Fig. 18C).

TREATMENT

Following preoperative management and medication, the infant was placed in a lateral decubitus position with the right side of the chest uppermost. Using an extrapleural surgical approach, the thorax was entered by removing the fourth rib and reflecting the parietal pleura from the thoracic wall until the posterior mediastinum was entered. The vagus nerve, upper and lower esophageal pouches, and the fistulous connection between the lower esophagus and trachea were identified. The connection between the lower esophageal pouch and the trachea was separated, and the stability of the suture line on the trachea was tested by submerging the latter in a saline solution and applying intratracheal pressure. The upper esophageal pouch was mobilized first, followed by the lower esophageal pouch. Any gap between the two portions of the esophagus was overcome by additional mobilization of the upper pouch due to its greater vascularity and thicker wall. The upper portion of the lower esophagus was excised to produce a larger lumen and increased blood supply, and a separate mucosal and muscularis esophageal anastomosis was made to prevent recurrence of a fistula. The surgery was well tolerated and postoperative feedings of glucose and saline by dropper were continued for three days. On the fourth day an aqueous contrast swallow was performed to determine any leakage of the anastomosis and the passage of material into the stomach. With no evidence of leakage, oral feeding of glucose and boiled skim milk was initiated, with full caloric intake reached by the fourteenth postoperative day. Bottle feeding began the third week. Except for several dilatations of the anastomotic site, the patient made an uneventful recovery and at 2 years of age is well developed and free from any symptoms.

DISCUSSION

The first known description of congenital esophageal atresia with tracheoesophageal fistula was given by Thomas Gibson[12] in 1697. Further cases,[26, 62] were not noted until the 1800's and, until the late nineteenth century, interest was concentrated on the embryology and pathology with only sporadic attempts at treatment.[16, 34] In 1929 Vogt[42] presented a classification, still in use, of this defect which affects approximately 1 of 2,500 to 3,000 newborn infants.[35] Lanman[21] and Shaw[38] were the first to attempt a primary anastomosis of the two esophageal segments and in 1943 Haight and Towsley[15] were the first to succeed with a primary extrapleural esophageal anastomosis. A multiple-stage repair was initiated by Ladd[20] and Leven[22] and in 1962 Holder, Mc-

Donald, and Wooley[17] reported increased success for critically ill and premature infants using a staged repair.

Embryology

Explain the occurrence of this malformation embryologically. Differentiation of the esophagus and the trachea occurs between the third and sixth week of gestation, at which time this malformation originates. In the 3-week-old embryo, a groove, the tracheobronchial sulcus, develops on the floor of the foregut between the thyroid anlage cranially and the liver primordium caudally. The epithelium of the foregut is one to two cells thick, whereas in the region of the tracheobronchial sulcus it is three or more cells thick. The ventral wall of the sulcus thickens as a result of epithelial proliferation and lateral ridges form in a coronal plane. These ridges, upon eventual fusion, separate the caudal portion of the developing trachea from the esophagus. This fusion gradually extends cranially. As a result of incongruence in growth between epithelial and mesodermal components, the epithelial cells are abundant and tend to fill the esophageal lumen. Later, as the mesoderm increases, the epithelium arranges itself along the wall of the esophagus and the width of the lumen increases. It is probable that this epithelial proliferation occurs throughout the digestive tract, especially in the esophagus, duodenum, colon, and rectum.

The various theories proposed for the occurrence of esophageal atresia and tracheoesophageal fistula include: (1) inflammation and ulceration of the developing esophagus and trachea; (2) deficiency of material for the complete construction of two separate systems; (3) relative pressure changes; (4) malposition of the large vessels, especially the right subclavian artery; (5) vascular insufficiency to the area; and (6) deviation of the normal course of the tracheobronchial sulcus, which begins at the site of tracheal bifurcation and eventually separates the trachea from the esophagus.

Classification

Classify the various forms of congenital atresia of the esophagus (Fig. 18). Numerous classifications exist; the best known is that of Vogt.[42]

Type 1. Congenital agenesis of the esophagus.
Type 2. Atresia of the esophagus without fistula.
Type 3. a. Atresia of the esophagus with tracheoesophageal fistula of the proximal esophagus.

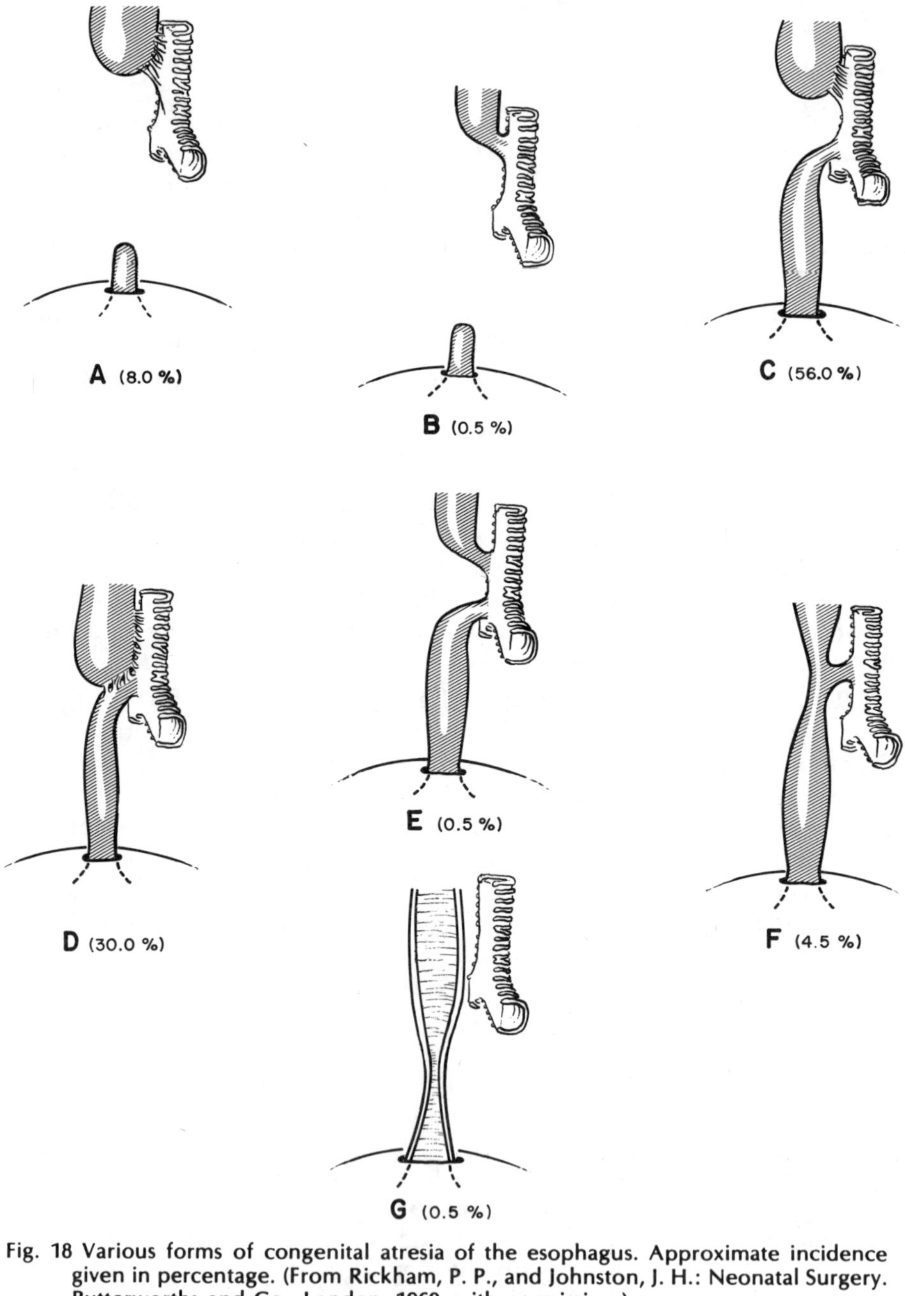

Fig. 18 Various forms of congenital atresia of the esophagus. Approximate incidence given in percentage. (From Rickham, P. P., and Johnston, J. H.: Neonatal Surgery. Butterworths and Co., London, 1969, with permission.)

b. Atresia of the proximal esophagus with tracheoesophageal fistula of the distal esophagus.
c. Atresia of the esophagus with tracheoesophageal fistula of both the proximal and the distal esophagus.

The Vogt classification was modified by Ladd[20] and by Roberts[35] into that currently used:

1. Atresia of the esophagus without fistula.
2. Atresia of the esophagus with fistula between the proximal segment of the esophagus and the trachea.
3. Atresia of the esophagus with fistula between the distal segment of the esophagus and the trachea.
   a. Proximal and distal esophageal segments are widely separated.
   b. Proximal and distal esophageal segments are close together or overlapping.
4. No atresia of the esophagus. "H"-type tracheoesophageal fistula.
5. Stenosis of the esophagus.

Surgical Approach

What are the advantages and disadvantages of the extrapleural and transpleural surgical approaches? The extrapleural approach presents less trauma to the lung, which remains behind the parietal pleura. If esophageal anastomotic leakage occurs, the infection remains localized and can be readily drained. The main disadvantages are the long operating time, limited exposure, frequent injury to the parietal pleura, and the reduced lateral motion postoperatively due to resection of a rib.

The transpleural approach allows a shorter operating time, wide field of exposure, and the use of mediastinal material for anastomotic suturing. The great disadvantage is that if an anastomotic leakage occurs, empyema and death result in a high percentage of the cases; also there is unavoidable trauma to the lung, which may produce atelectasis and subsequent pneumonitis.

Postoperative Complications

What factors may cause a complicated postoperative course? Pulmonary complications–i.e., atelectasis, pneumonia, pneumothorax, and empyema–anastomotic leakage, or stricture account for most postoperative complications.

Associated Malformations

What major malformations are frequently associated with esophageal atresia and tracheoesophageal fistula? Atresia of other portions of the gastrointestinal tract, imperforate anus, lesions of the aortic arch, and polycystic kidney are closely correlated with this congenital defect.

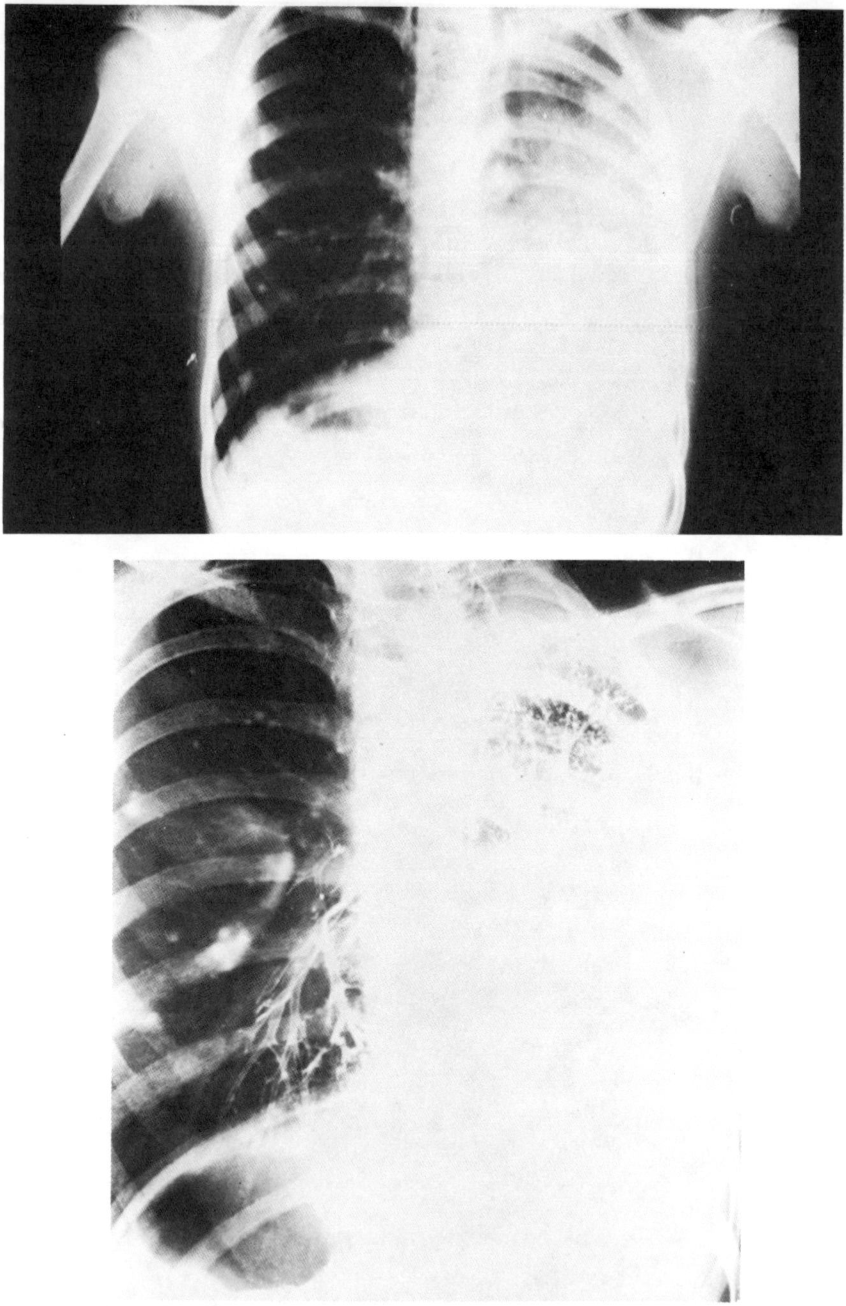

Above, X-ray picture showing agenesis of the left lung; *Below,* Bronchogram of same infant showing little evidence of a left bronchus.

12

Agenesis of the Lung

A male infant, 8 months old, was admitted to Children's Hospital for wheezing and difficulty in breathing. Since birth these symptoms had gradually increased, but there was no cyanosis and prior to hospital admission he was treated for pneumonia. Physical examination showed the chest to be asymmetric. The left side was smaller and demonstrated diminished movement, dullness on percussion, and absence of breath sounds.

X-ray and fluoroscopy showed the entire left lung field to be opaque, while the right lung showed marked overexpansion. The mediastinum was shifted to the left side, which revealed narrower intercostal spaces than on the right. Thirteen ribs were present on each side.

A bronchogram showed no evidence of a left bronchus or its vestige.

The infant was bronchoscoped under local anesthesia. The larynx was rotated toward the left, but the vocal folds moved normally. The trachea was displaced to the left; the left primary bronchus was totally obstructed about 1 cm. below the carina, whereas the right bronchial tree was normal. No other malformations were noted.

DIAGNOSIS

Agenesis of the left lung.

FURTHER COURSE

In the absence of other malformations, unilateral agenesis is compatible with normal existence, but the prevention or care of respiratory infections is mandatory. The infant was discharged under pediatric care for periodic checkups. Over the next four years he showed normal development and his breathing difficulties subsided.

DISCUSSION

Complete agenesis of pulmonary tissue is extremely rare except in acephalic monsters. Unilateral agenesis is not uncommon; it is found more often in males, and the left lung is affected twice as often as the right. Morgagni described agenesis of the lung in 1762,[25] and in 1885 Munchmeyer[43] was the first to attempt diagnosis of unilateral pulmonary agenesis, later verified at necropsy. Using clinical evidence, von Ecker[8] made the first correct diagnosis. Bonniger (1928)[3] made the diagnosis based on bronchography, and Stokes and Brown (1940)[40] confirmed the diagnosis bronchoscopically.

Embryology

Discuss normal lung development. At three and one-half weeks of gestation (Fig. 19A), the respiratory system (larynx, trachea, lungs)

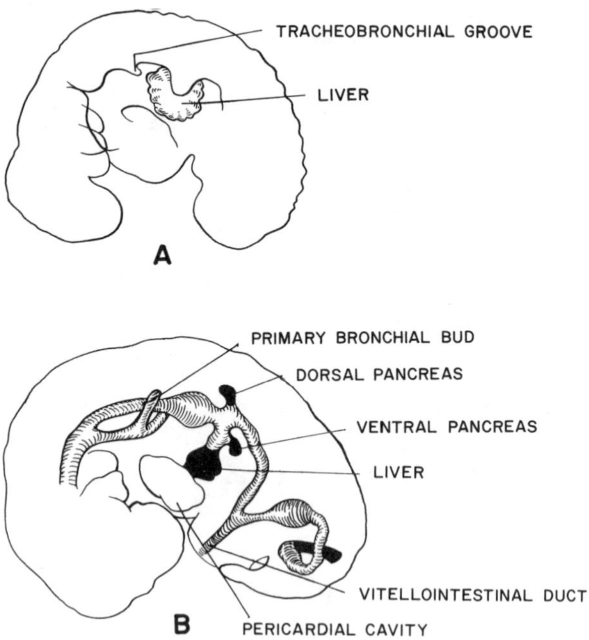

Fig. 19 Developing respiratory system: *A*, 4-week-old embryo; *B*, 5-week-old embryo.

begins to develop as a tracheobronchial groove, an endodermal outgrowth from the floor of the primitive foregut. This groove may represent fused pharyngeal pouches. The mesenchymal constituents of the respiratory system arise from splanchnic mesoderm on the ventral surface of the foregut into which the respiratory diverticulum and its subsequent subdivisions extend.

At four weeks (Fig. 19B), the caudal end of the tracheobronchial groove grows ventral to the foregut, forming an independent tracheal tube that is dilated at its distal end to form the primary bronchial buds. Soon the bronchial buds become asymmetrical, the left one lying more transversely than the right. Early branching of the primary bronchi tends to be monopodial (Fig. 20); i.e., a branch is formed on one side of the bronchi while the main stem continues to grow beyond this point without essential change of direction. Branching after the pri-

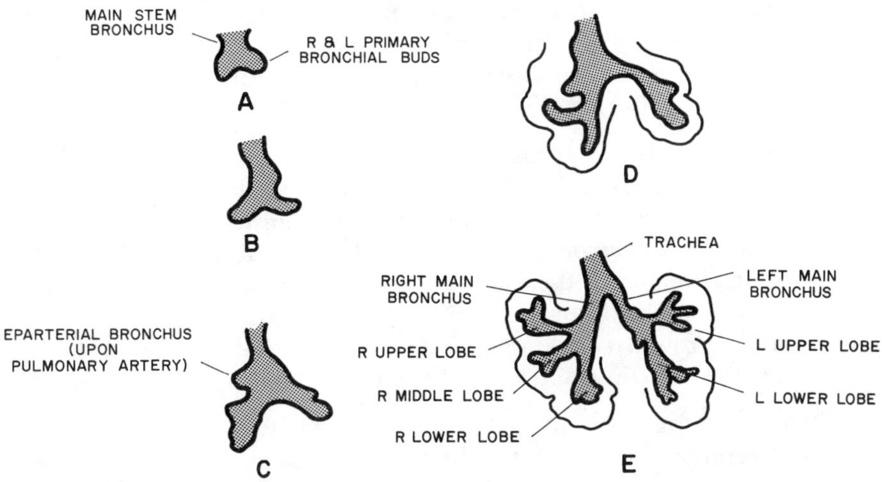

Fig. 20 Bronchial tree development: *A*, 28 days; *B*, 30 days; *C*, 32 days; *D*, 33 days; *E*, 35 days.

mary bronchi are established is more likely to be dichotomous, i.e., bifurcation into two symmetrically placed branches, neither of which extends in the direction of original growth. At this stage (Fig. 20E), the right lung bud possesses three bronchial tubes, while the left has two. These endodermal tubes, together with the lung bud mesenchyme that surrounds them, give origin to the lobes of the adult lungs. The part of the foregut dorsal to the trachea forms the esophagus and is probably a result of interstitial growth, since the cranial and caudal ends of the esophagus are fixed in position.

Once the primary bronchi are established, four stages of bronchial tree development occur:

1. *Glandular*—to the fourth month of gestation, during which time the remaining divisions of the bronchial tree are formed and respiratory-like movements ensue.
2. *Canalicular*—from the fourth to the sixth month, the respiratory portion is delineated. The lung mesenchyme becomes highly vascular and capillary loops surround the developing alveoli.
3. *Alveolar*—from six months to term, branching continues until 24 generations of alveoli, which is the characteristic number in the adult lung, are formed. By the seventh month, capillaries surrounding the alveoli are sufficiently developed to support extrauterine existence. As additional alveolar sacs form, the tissue separating the blood of the pulmonary circuit from the alveolar air becomes sufficiently thin to readily permit exchange of O_2 and CO_2.
4. *Birth*—the terminal bronchioles continue to divide postnatally to form six or seven additional generations; this can occupy the first eight years of life. At birth the initiation of respiration dilates and expands the alveoli. This expansion begins in the central region of the lung and at approximately three days postnatal the alveoli of the peripheral areas are dilated. Surfactant, a lipoprotein compound, coats the epithelial surface of the alveoli and contributes to the stabilization of alveolar surface tension.

Fetal aspiration of amniotic fluid into the lungs may occur, but is thought to result only from asphyxia or a similar condition. The fluid normally present in the fetal alveoli and tracheobronchial tree is not aspirated amniotic fluid, but an ultrafiltrate of plasma.

Causes of Lung Agenesis

Several factors are proposed as the cause of lung agenesis. They include disturbances in: (1) genetics; (2) fetal nutrition; and (3) developmental mechanics. Additional causes may include interference with fetal blood supply and disturbances in closure of pleuroperitoneal communications.

Differential Diagnosis

What lung malformations or disease can interfere with the diagnosis of lung agenesis? Atelectasis, bronchiectasis, postpneumonectomy chest, fibrothorax, and blockage of a main bronchus by tumor give symptoms similar to those of agenesis. These may be eliminated using bronchoscopy and bronchogram.

Classification

Two sets of classifications have been used to describe lung malformations:

1. Schneider's Classification[37]
   a. Agenesis—complete absence of a lung, bronchus, or vascular supply
   b. Aplasia—presence of bronchial rudiments without alveolar tissue
   c. Hypoplasia—deficient growth of regularly differentiated pulmonary components
2. Minetto's Classification[27]
   a. Bilateral pulmonary agenesis
   b. Unilateral pulmonary agenesis
   c. Pulmonary aplasia
   d. Pulmonary hypoplasia
   e. Lobar aplasia

Associated Malformations

What malformations are most commonly associated with lung agenesis? Narrowed trachea, extracartilaginous rings, supernumerary bronchi of the normal lung, absence of pleura, tracheoesophageal fistula, and esophageal stenosis are most frequently found with lung agenesis. In addition, defects of almost every tissue and organ have been reported in conjunction with agenesis.

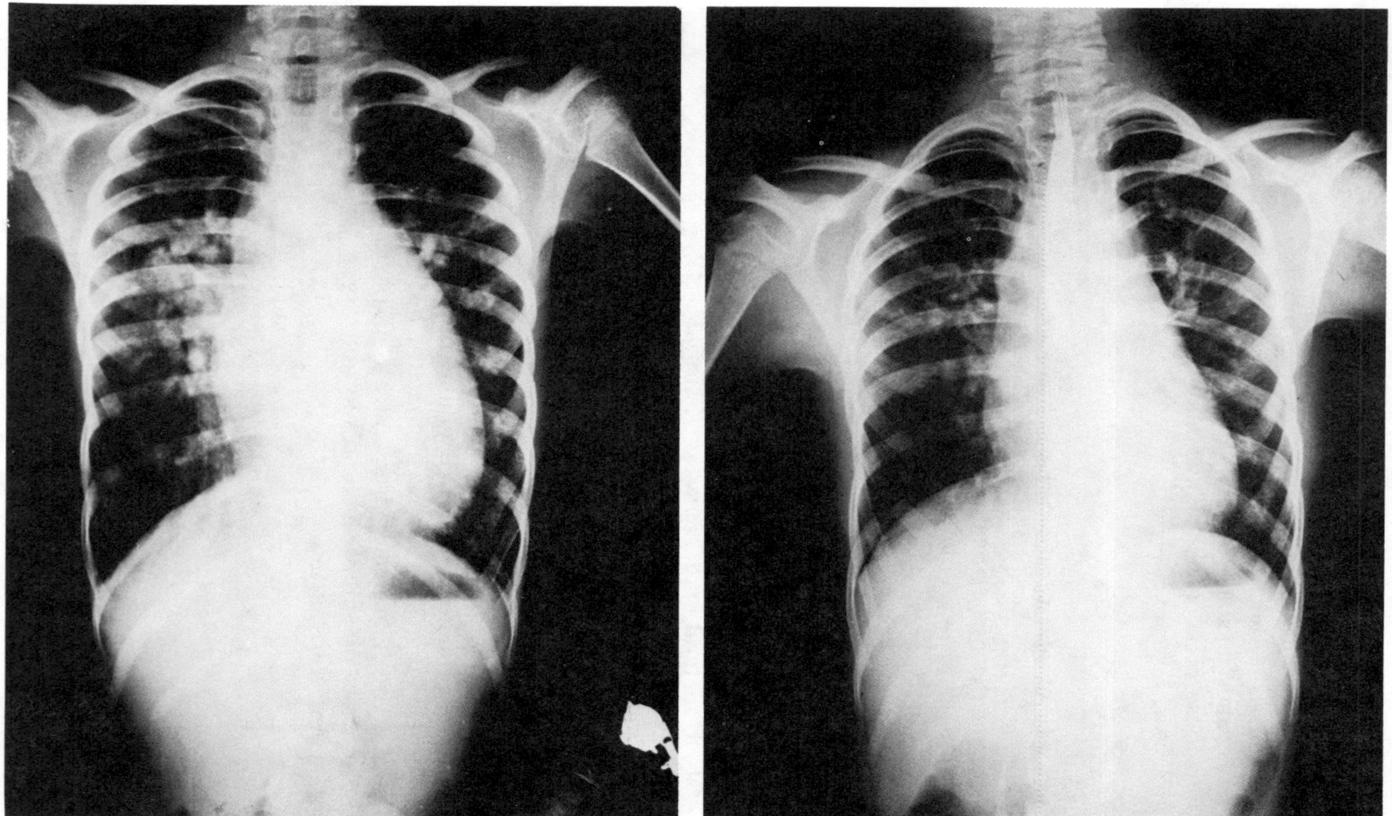

Left, X-ray study of a patent ductus arteriosus showing slight left ventricular hypertrophy with an increase of lung vascularity; *Right,* One year after closure of the patent ductus arteriosus.

13

Patent Ductus Arteriosus

An 18-month-old acyanotic female is admitted to Children's Hospital for cardiac catherization. Her physical development (height and weight) is retarded; mental development is normal. Recently, a heart murmur was detected during treatment for pneumonia. Upon admission, a continuous characteristic ("machinery")[11] murmur, accentuated during systole, is heard in the second intercostal space to the left of the sternum. A thrill is present over the precordium. Radiologic examination reveals a slight left ventricular hypertrophy with an increase of lung vascularity. Cardiac catheterization performed on the fifth day following admission gives direct measurements of blood flow and confirms the diagnosis of patent ductus arteriosus with a left-to-right shunt. Surgery is indicated.

Family history reveals that the mother had rubella in the first few months of pregnancy.

DIAGNOSIS

A medium-sized patent ductus arteriosus (PDA).

TREATMENT

With the infant under general anesthesia, a posterolateral incision was made in the fourth interspace and the mediastinal pleura separated between the phrenic and vagus nerves. The vagus was retracted to expose the recurrent laryngeal nerve where it passes behind the ductus. Dissection began close to the aorta, removing the pericardium back to the pulmonary artery. Potts clamps[32] were applied to the two ends of the ductus, which was divided and the ends oversewn. The thoracic wall was closed, recovery was uneventful, and on the tenth postoperative day examination revealed the absence of a precordial thrill and murmur; roentgenologic evidence showed a decreasing volume of the heart shadow and of lung vascularity.

DISCUSSION

Although originally described by Galen[10] as a short vessel uniting the aorta and the pulmonary artery, the ductus arteriosus is erroneously associated with the name of Botallo (seventeenth century), because in 1766 von Haller[10] translated the published works of Botallo and inserted one of his own drawings depicting a connecting vessel between the aorta and the pulmonary artery. Gibson (1900)[11] described the now-classic "machinery" murmur, and in 1907 Munro[28] suggested a surgical approach and operative closure of the patent ductus. Graybiel, Strieder, and Boyer (1938)[13] attempted closure of an infected patent ductus,

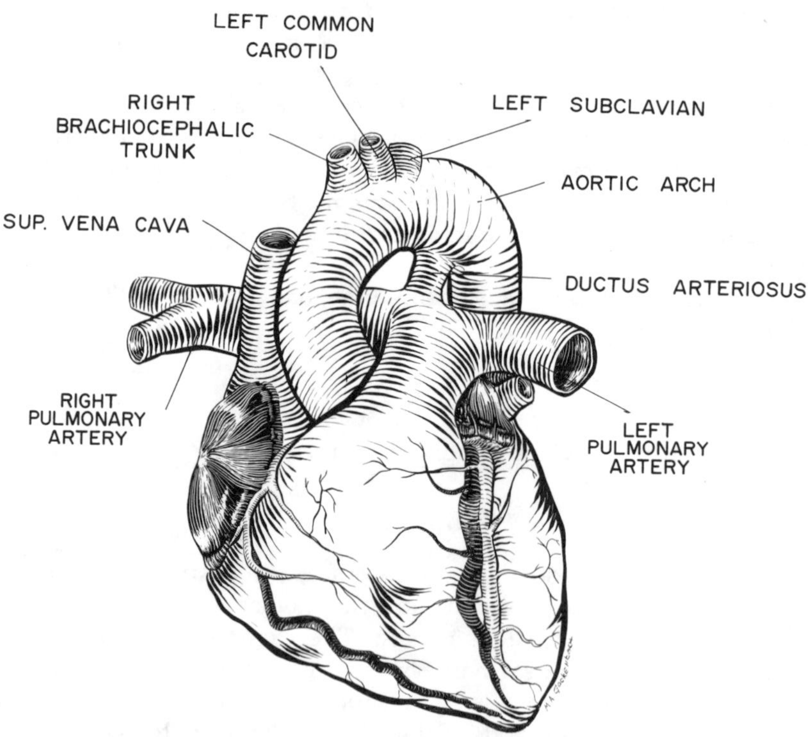

Fig. 21 Anatomical relationships of a ductus arteriosus in the newborn.

but the infant succumbed postoperatively. In 1939 Gross and Hubbard[14] were the first to successfully close a patent ductus.

Definition

Define ductus arteriosus. The ductus arteriosus (Fig. 21), the distal portion of the sixth left aortic arch, is a thick-walled arterial channel connecting the left pulmonary artery with the aorta immediately distal to the origin of the left subclavian artery. Patent ductus arteriosus occurs in one individual out of every 3,850 in the general population.[18] The ductus arteriosus is a channel through which blood bypasses the lungs during the fetal period and it closes spontaneously shortly after birth. The mechanism of closure is debated, but probably at the onset of respiration, a drop in the pulmonary vascular resistance results in a pronounced constriction of the ductus. Closure of the ductus is completed by fragmentation of the internal elastic lamina, allowing the smooth muscle to project into the lumen. Fibrosis then converts the ductus arteriosus into the ligamentum arteriosum.

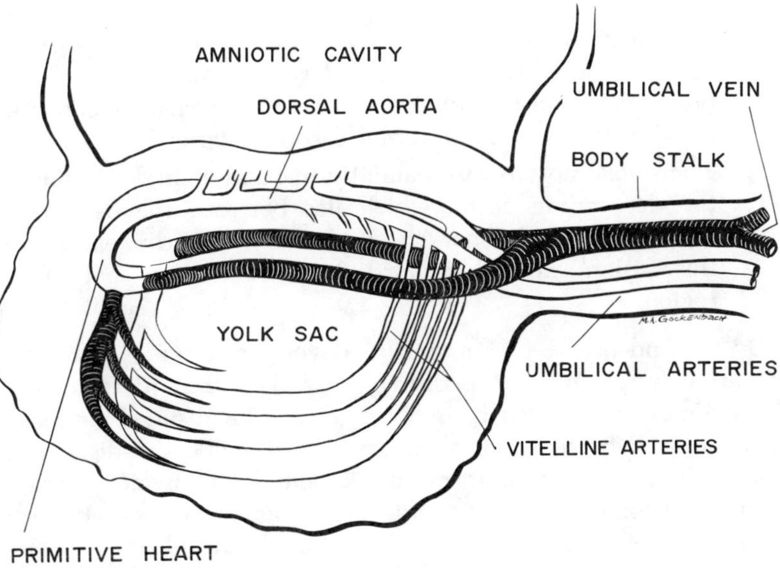

Fig. 22 Aortic arch development: 2.5-week-old embryo.

Embryology

Discuss embryologically the development and subsequent differentiation of the aortic arch arteries (aortic arches) and the aorta, and explain the occurrence of the ductus arteriosus. The primary embryonic arteries are the right and left primitive aorta. They are identified in early somite embryos (Fig. 22) as two vessels that appear as continuations of the two endocardial tubes. Each passes from the region of the floor of the foregut around the pharynx and to the dorsal aspect of the gut as the dorsal aorta. Each primitive aorta is divided into three portions: (1) ascending, (2) pharyngeal arch, and (3) descending. Fusion of the two endocardial tubes (Fig. 23A) reaches the anterior limit of the pericardial cavity and extends craniad to include the cardiac extremities of the two ascending portions. The latter extend to the bulbus cordis, which continues cranially beyond the pericardium as a single midline vessel, the truncus arteriosus (Fig. 23B). The truncus becomes dilated (three and one-half weeks) and is known as the aortic sac. The right and left first aortic arches arise from this sac. As development proceeds, successive pairs of aortic arches differentiate caudally in association with each forming pharyngeal arch and serve as communication channels between the aortic sac and the dorsal aortae. Shortly after the appearance of the aortic sac and the first aortic arch, the dorsal aortae undergo two changes:

1. Each aorta grows in a cranial direction beyond where the first aortic arch joins it, and in close relation to the developing brain plate. In the adult these cranial extensions constitute the distal portions of the internal carotid arteries.
2. The dorsal aortae fuse caudally to form a single midline vessel, the descending aorta, which lies between the developing gut ventrally and the notochord dorsally. As the embryo grows, this fusion extends forward, but not to the pharyngeal arch region.

By the end of the first month of development (Fig. 23B), the first aortic arch disappears as the third arch fully differentiates; the second arch involutes as the fourth arch matures. The distal remnants of the first aortic arch become the maxillary artery, while the distal remnants of the second aortic arch form the stapedial and hyoid arteries. The proximal portion of the first and second aortic arches attach to the third aortic arch to form part of the external carotid artery. The third aortic arch from its origin at the aortic sac to the attachment of the first and second aortic arches forms the common carotid artery. The

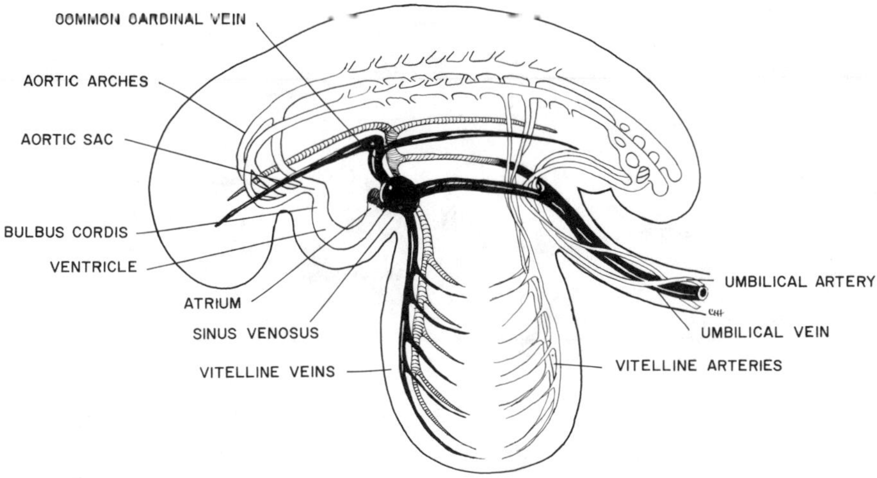

A

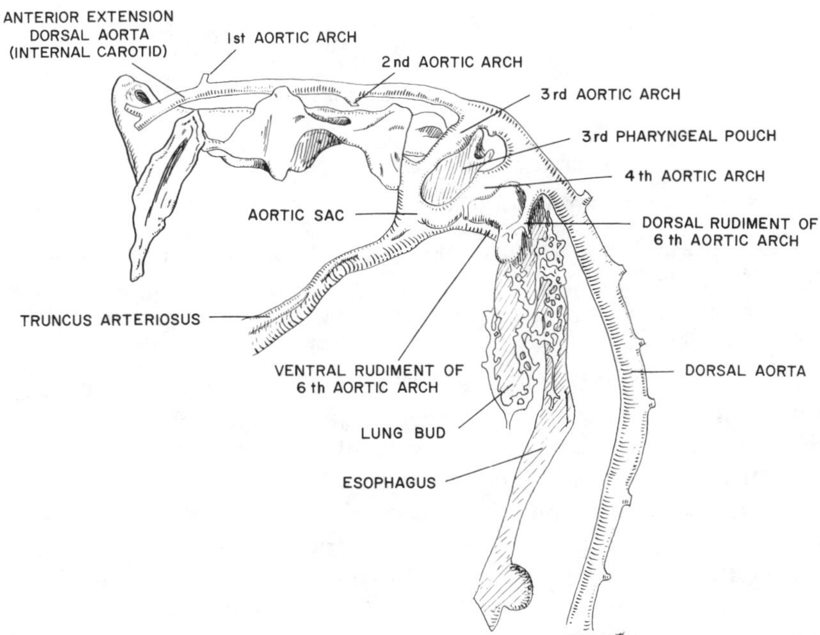

B

Fig. 23 Sagittal view of aortic arch development: *A*, 4-week-old embryo; *B* and *C*, Enlargements of area outlined in *A*. *B*, 4-week-old embryo; *C*, 6-week-old embryo.

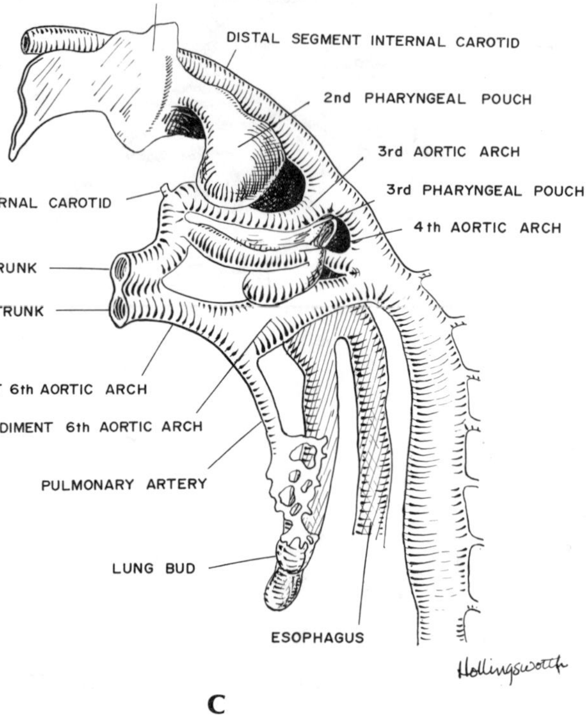

Fig. 23 (continued. See page 99 for legend.)

fifth aortic arch exists transiently. The sixth aortic arch is divided into a ventral rudiment that forms the pulmonary artery, and a dorsal rudiment that arises from the dorsal aorta and joins the ventral rudiment as the ductus arteriosus.

By the sixth week (Fig. 23C), primarily because of changes in vascular hemodynamics, the arterial system completes its formation and undergoes modifications to the adult pattern. The third, fourth, and sixth aortic arches are developed, and the bulbus cordis becomes longitudinally divided into two parts by the growth of the spiral bulbar septum. The distal part of the septum extends into the aortic sac such that blood which enters the bulbus from the right ventricle empties into the dorsal portion of the aortic sac and into the sixth aortic arch only; blood from the left ventricle empties into the ventral portion of the aortic sac and is distributed to the third and fourth aortic arches. Disappearance of the cranial portion of the right dorsal aorta between its attachment to the left dorsal aorta and to the right fourth aortic arch subsequently results in the loss of the dorsal rudiment of the right sixth aortic arch beyond the origin of the pulmonary artery. This dorsal rudiment corresponds to the ductus arteriosus of the left side. The fourth aortic arch persists bilaterally; on the left it joins with the aortic sac and the left horn of the aortic sac to form the arch of the aorta up to the point of attachment of the seventh segmental artery, the latter which forming the left subclavian artery. On the right, the horn of the aortic sac becomes the brachiocephalic artery, which forms the stem of the right common carotid and subclavian arteries; the remainder of the subclavian is formed by the right fourth aortic arch (Fig. 24).

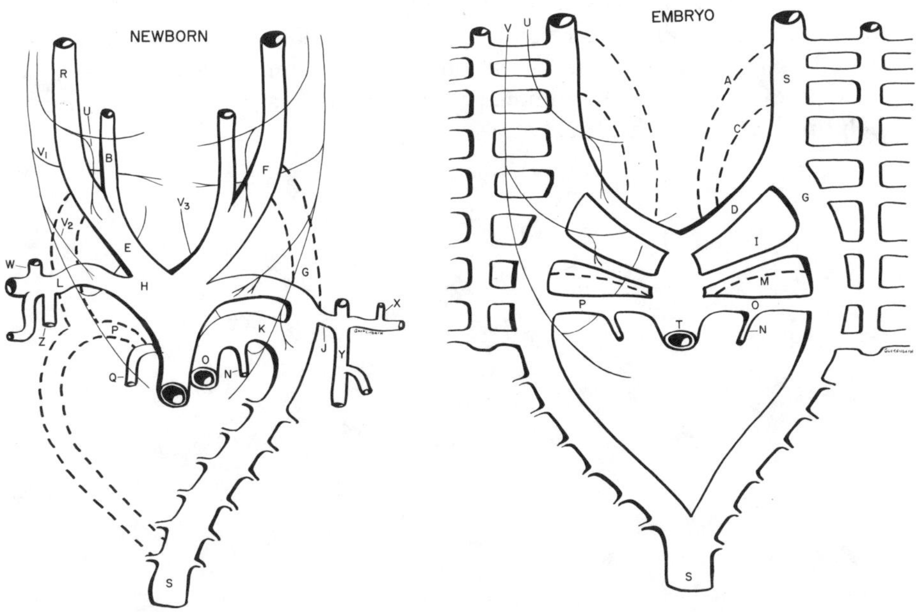

Fig. 24 Comparison of aortic arch system (frontal view): *A*, 1st arch artery (maxillary); *B*, External carotid; *C*, 2nd arch artery (stapedial, hyoid); *D*, 3rd arch artery; *E*, Common carotid (3rd arch); *F*, First part internal carotid artery (3rd arch); *G*, Ductus caroticus (dorsal aorta between 3rd and 4th arch); *H*, Brachiocephalic (aortic sac); *I*, 4th arch artery; *J*, 4th arch at attachment of 7th intersegmental (subclavian); *K*, Ductus arteriosus (dorsal rudiment of left 6th arch); *L*, Right subclavian; *M*, Transient (?) 5th arch artery; *N*, Ventral rudiment 6th arch artery (left pulmonary); *O*, Pulmonary trunk; *P*, Dorsal rudiment of 6th arch on right; *Q*, Right pulmonary artery; *R*, Head segment of internal carotid; *S*, Fused dorsal aorta; *T*, Aortic sac; *U*, IX nerve; *V*, Superior laryngeal n. (X); V , (X); V , Recurrent laryngeal (X; *W*, Vertebral (longitudinal anastomosis of dorsolateral intersegmentals); *X*, Thyrocervical trunk; *Y*, Costocervical trunk; *Z*, Internal thoracic.

Differential Diagnosis

What heart or arterial malformations may be confused with a patent ductus arteriosus? A PDA is most frequently confused with aorticopulmonary window (congenital communication between the ascending aorta and the stem of the pulmonary artery directly above the semilunar valves, resembling hemodynamically and clinically a PDA). Other conditions that must be distinguished are venous hum, pulmonary arteriovenous fistulas, ventricular septal defect, aortic insufficiency, ruptured sinus of Valsalva, and coronary arteriovenous fistula.

Associated Malformations

What malformations are frequently associated with a PDA? Coarctation of the aorta and ventricular septal defect are frequently associated malformations. In conditions such as pulmonary stenosis, tetralogy of Fallot, and pulmonary or tricuspid atresia, PDA may be helpful in keeping the child alive.

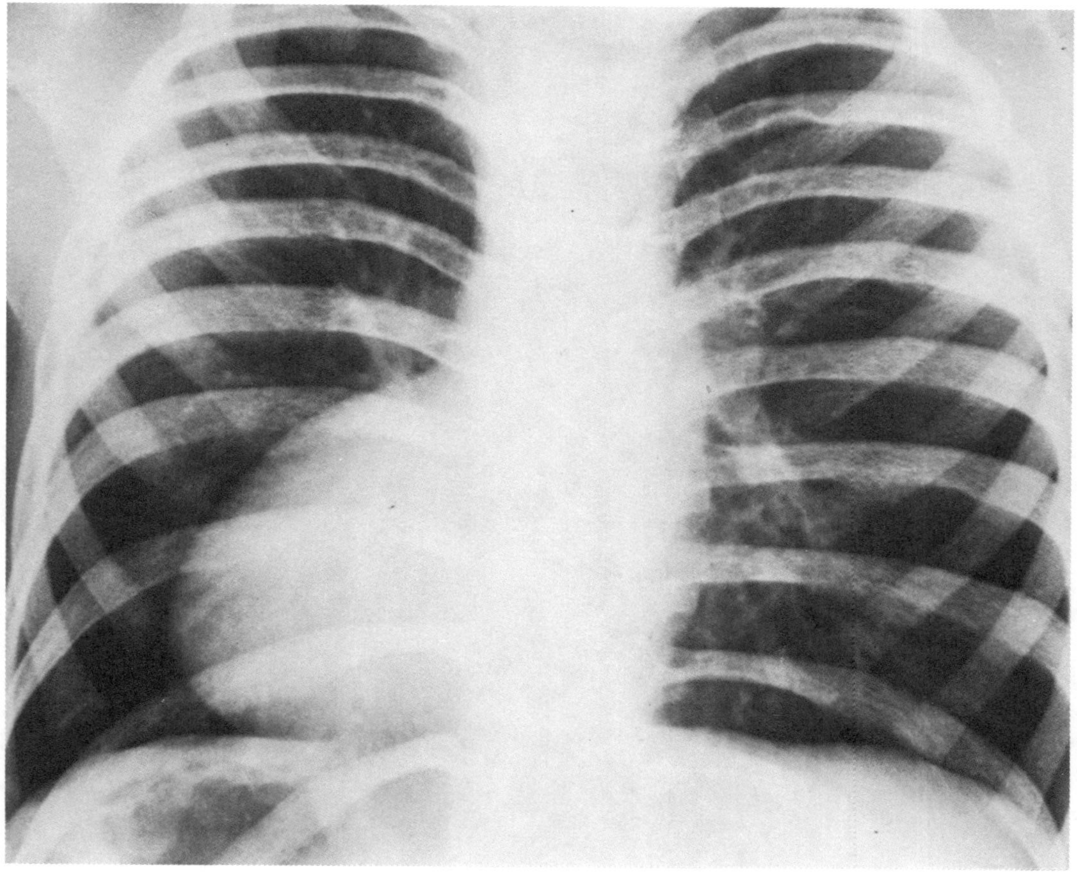

Couer en sabot of the tetralogy of Fallot.

14

Tetralogy of Fallot

A 2-year-old boy was admitted to the hospital for evaluation of a heart murmur previously detected at birth. He was less active than other children his age, but, although overexertion was followed by frequent squatting and cyanosis of the lips and nails, there was no history of loss of consciousness. Initial examination revealed a thin, physically retarded, cyanotic child with no respiratory difficulty. Moderate clubbing of fingers and cyanosis of the nail beds was noted. A harsh systolic murmur was maximal over the mid-precordial area; no thrill was felt. The first heart sound was normal while the second was single, distinct, and loud. The lungs were clear. Roentgenologic study showed a normal-sized heart dominated by a boot-shaped (couer en sabot) right ventricular outflow-tract. Peripheral vascular shadows in the lungs were less prominent than normal. Electrocardiography indicated a right axis deviation of the heart, while angiocardiography showed a right-to-left shunt through a ventricular septal defect, early appearance of dye in the ascending aorta due to its overriding position, and diminished blood flow through the lung fields.

DIAGNOSIS

Tetralogy of Fallot by definition consists of a ventricular septal defect and pulmonic stenosis (valvular or infundibular) resulting in an overriding aorta and right ventricular hypertrophy (Fig. 25).

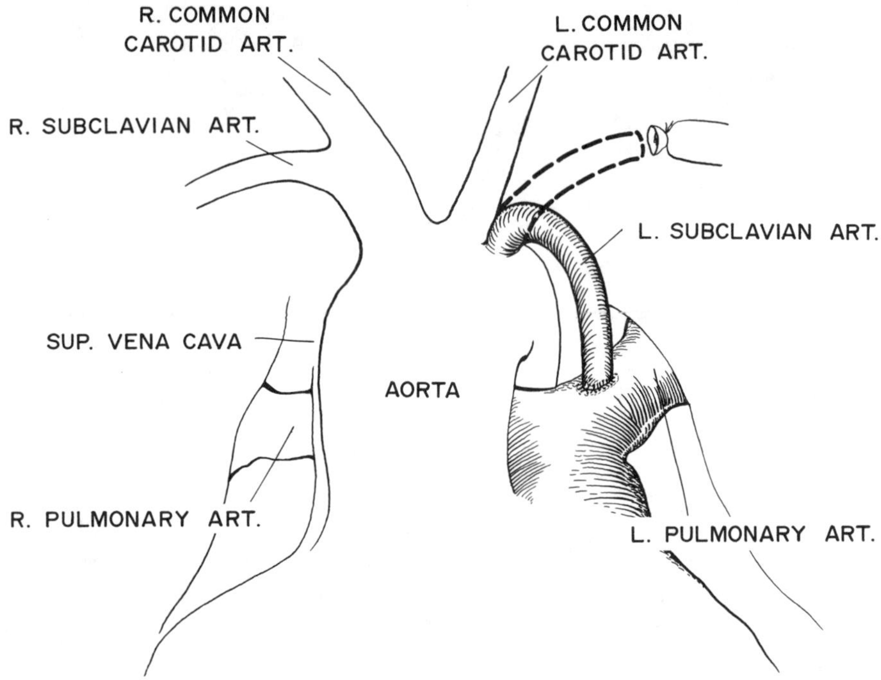

Fig. 25 Blalock-Taussig shunt.

TREATMENT

When the child was 2 years of age, a palliative Blalock-Taussig shunt (**Fig. 25**) was performed by anastomosis of the divided proximal end of the left subclavian artery to the side of the left pulmonary artery, resulting in increased collateral circulation to the lungs, thereby relieving the cyanosis. Open intracardiac repair was performed when the child was seven years old, with closure of the palliative collateral channel, relief of the pulmonary outflow-tract obstruction, and complete closure of the ventricular septal defect. The patient had an uncomplicated recovery and is acyanotic and asymptomatic one year later.

DISCUSSION

Six out of 1,000 live births are affected by anomalies of the heart and great vessels.[44] Stensen in 1671[39] first described the anatomy that was to become known as tetralogy of Fallot. In 1888[9] Fallot defined the four basic components comprising the tetralogy. Diagnosis and palliative treatment advanced under Blalock and Taussig (1945)[2] and under Potts, Smith, and Gibson (1946).[33] In 1955 Lillehei and coworkers[23] and Kirklin and colleagues[19] began total open intracardiac correction of tetralogy of Fallot utilizing a cardiopulmonary bypass.

Embryology

Discuss developmentally the defects that characterize tetralogy of Fallot. In the late presomite embryo, the angiogenic tissue from which the primordium of the heart will develop first appears as scattered groups of cells arising in the mesenchyme of the anterior margin of the embryonic disk cranial to the neural plate; this is the cardiogenic area. These groups of angiogenic cells soon fuse into a plexus of endothelial vessels which unite to form right and left endocardial heart tubes; the latter gradually merge to establish a single median endocardial tube. At four weeks of gestation, the primitive heart is not divided into separate chambers, but undergoes differential expansion that results in several dilatations separated by grooves. From caudal to cranial, the dilatations are the sinus venosus, the atrium, the ventricle, and the truncus arteriosus. Subsequently, a major change occurs with the partitioning of the heart tube by the formation of the cardiac septa.

Early in the second month (Fig. 26A), indication of the division of the primitive ventricle into right and left chambers occurs with the development of the primary muscular partition of the interventricular septum. This crescent-shaped septum appears at the apex of the ventricular bend and is directed toward the artioventricular canal cushions (dorsal and ventral endocardinal cushions which separate the atria from the ventricles). An interventricular foramen persists for a short time between the interventricular septum and the canal cushions; this foramen is closed by material derived from several sources (i.e., conus ridges, endocardial cushions, and the crest of the original interventricular septum).

Truncus ridges between the roots of the fourth and sixth aortic arches pursue a spiral course through the truncus arteriosus toward the ventricles, and form a complete partition separating an aortic channel leading into the fourth aortic arches from a pulmonary channel leading into the sixth arches (Fig. 26B). The ascending aorta and the main pulmonary trunk (Fig. 26B) twist around each other as they emerge from the ventricles, thereby bringing the aortic and pulmonary channels into contact with the left and right ventricles, respectively. The level at which the aortic and pulmonary valves develop is regarded as the

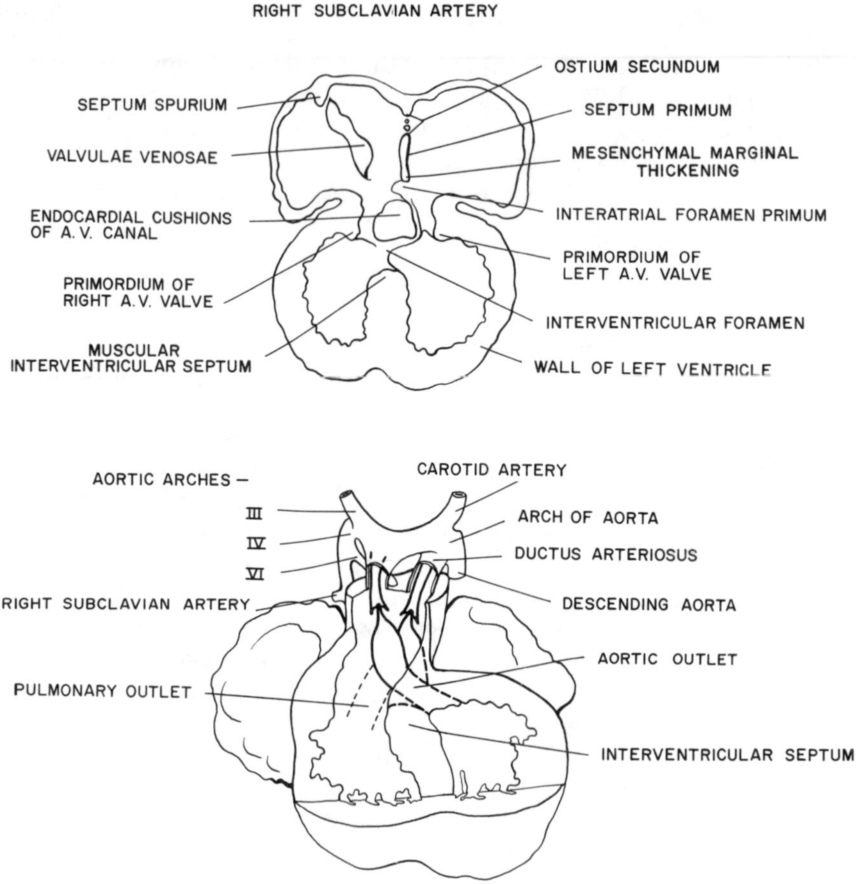

Fig. 26 Internal development of the heart: *A*, 5-week-old embryo; *B*, 3-month-old fetus.

line of demarcation between the truncus arteriosus and the tapering ventricular outlet; this is the conus arteriosus (infundibulum). On the ventricular side of the aortic and pulmonary valves, the continuation of the truncus ridges form the conus ridges. The conus ridges unite with the crest of the interventricular septum, thereby reducing the size of the interventricular foramen. Local enlargements of the endocardial cushions and of the connective tissue that caps the crest of the muscular portion of the interventricular septum complete the closure of the interventricular foramen. At the site of final closure, the interventricular septum gradually becomes a thin fibrous sheet called the membranous portion (Fig. 28E). The tetralogy of Fallot results from underdevelopment of the distal portion of the pulmonary conus ridge and is associated with inadequate posterior-superior growth of the muscular portion of the interventricular septum and incomplete formation of the membranous portion. Thus, the interventricular foramen is retained and accompanied by infundibular stenosis. With incomplete formation of the ventricular septum, the base of the aorta shifts to the right and overrides the muscular portion of the interventricular septum, allowing the aorta to communicate with both right and left ventricles. Infundibular stenosis results in right ventricular hypertrophy with a right-to-left shunt.

Palliative Procedures

What are the most common palliative procedures employed in children under 5 years of age? In the Blalock-Taussig shunt used in the

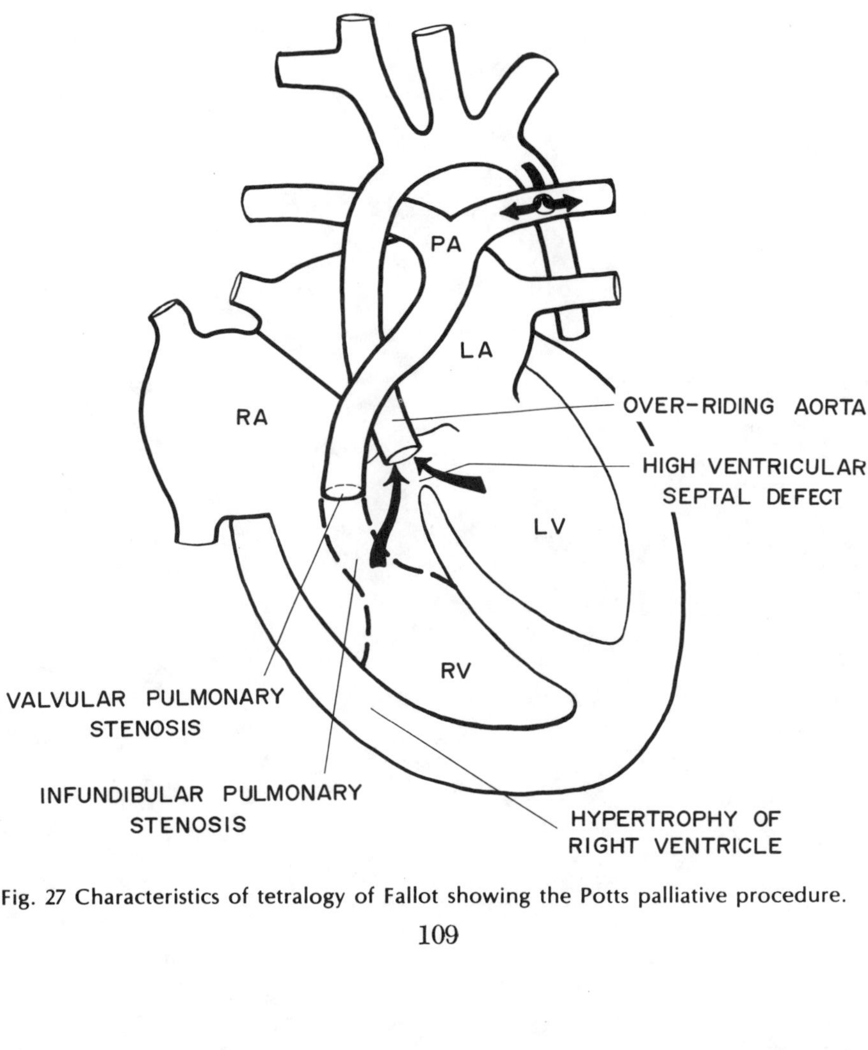

Fig. 27 Characteristics of tetralogy of Fallot showing the Potts palliative procedure.

present case, the Potts procedure (Fig. 27), which creates a stoma between the left pulmonary artery and the aorta, and the Brock[4] method of direct correction of the pulmonary obstruction are the principal palliative procedures. The main disadvantage of certain palliative methods is the difficulty with which they are dissected and closed at the time of total correction.

Associated Malformations

The most frequently associated major malformations are unilateral absence of the pulmonary artery and complete situs inversus. Minor malformations are an absence of the medial papillary muscle, alteration of the tricuspid valve, and occurrence of a right-sided aortic arch.

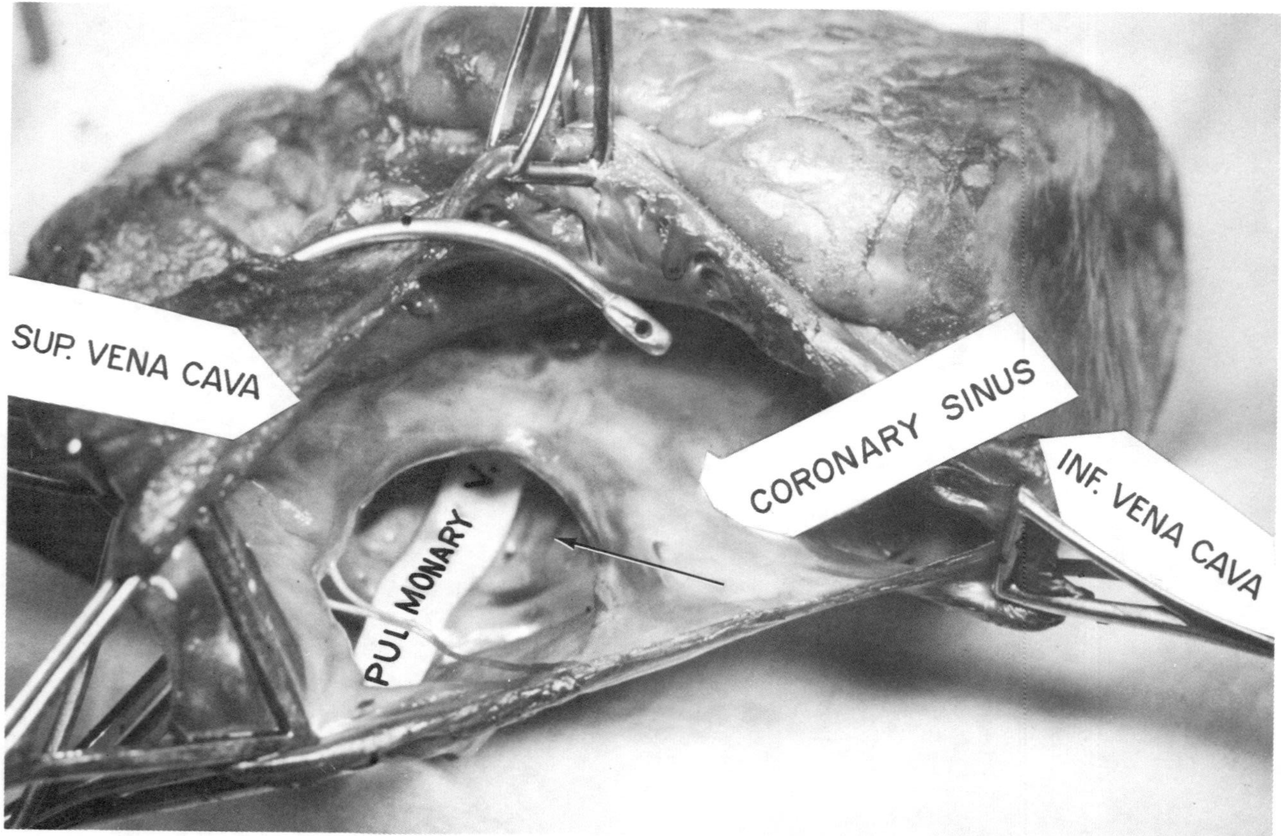

Atrial septal defect, *arrow* — ostium secundum type.

15

Atrial Septal Defect

A 6-year-old girl first demonstrated a cardiac murmur at 4 years of age. Growth and development were slightly retarded with some limitation of exercise tolerance. Cyanosis and clubbing were absent. The frequency of upper respiratory infections increased between the ages of 4 and 6.

EXAMINATION

Physical examination revealed a slight asymmetry of the thorax with prominence of the precordium. The heart was not markedly enlarged and no thrills were palpable. A systolic murmur was audible along the left sternal border with maximal intensity in the second left intercostal space. The electrocardiogram showed a right axis deviation and incomplete right bundle branch block. Fluoroscopic examination disclosed: (1) dilatation of the right heart and pulmonary arteries and veins; (2) pulsation of the pulmonary artery branches; and (3) increased vascularity of the lung fields. Cardiac catheterization verified the diagnosis of atrial septal defect and contributed pertinent information regarding the size and localization of the defect.

DIAGNOSIS

Atrial septal defect—ostium secundum type.

TREATMENT

Employing direct vision and hypothermia, the chest was opened through a split-sternal incision. Tapes were placed on the superior and inferior venae cavae, and the vessels were closed with rubber tourniquets and the heart emptied by massage. Air surrounding the heart was replaced by carbon dioxide during the cardiotomy. Then the atrium was opened by a longitudinal incision and the interior inspected, revealing a 4 cm. typical ostium secundum septal defect. With the defect visualized a suture was placed in its superior margin to clearly define the borders and to insure accurate placement of a suture at the inferior margin of the defect. The atrial septal defect then was closed by a double continuous running suture that included the wall of the left atrium to seal it effectively from the opening of the inferior vena cava. The caval tapes were removed and the atrial wall closed using running stitches. The initial incision was closed by interrupted wire sutures. The patient was asymptomatic 3 months postoperatively and cardiac catheterization revealed no left-to-right shunt at the atrial level.

DISCUSSION

Atrial septal defects occur in 7 per cent of all congenital cardiac malformations.[29] Von Ecker (1839)[8] and Cruveilhier (1852)[6] stated that dilatation and hypertrophy of the right heart, marked dilatation of the pulmonary artery and its branches, and the relatively small size of the left ventricle were distinguishing features of the heart having an interatrial communication. Nearly a century later, Cohn (1947)[5] closed artifically created atrial septal defects in dogs, while Murray (1948),[29] using a blind method of closure, was the first to operate on a human subject. Subsequently, several techniques of repair have been introduced.[1, 7, 30, 41] Today the open-heart operation with direct suture is most used, with some differences of opinion regarding which method of cardic bypass to utilize, the pump-oxygenator or hypothermia.

Embryology

Describe the partitioning of the primitive atrium. The common primitive atrium becomes divided into right and left chambers by the formation of two septa (Fig. 28E). Based on their sequential appearance, the septa are called septum primum and septum secundum.

In the second month of gestation (Fig. 28A), the septum primum, a crescentic ridge on the dorsocephalic part of the atrial wall, grows toward the atrioventricular canal. Concomitantly, two local thickenings (endocardial cushions) appear in the anterior and posterior walls of the atrioventricular canal. During the sixth week (Fig. 28B), the endocardial cushions fuse, forming a common mass dividing the atrioventricular canal into right and left channels.

Between the concave margin of septum primum and the endocardial cushions is a progressively diminished opening known as the ostium primum (interatrial foramen primum). As septum primum fuses with the endocardial cushions, ostium primum is obliterated; at the same time a new opening appears in the septum primum—the ostium secundum (interatrial foramen secundum).

In the seventh week (Fig. 28C), septum secundum forms in the roof of the atrium and to the right of septum primum. Like septum primum, it is crescentic in shape, but never forms a complete partition in the atrium. Septum secundum grows downward following the base of septum primum, and its concave margin overlaps the ostium secundum, leaving an oval aperture known as the foramen ovale (Fig. 28D). The margin of septum secundum constitutes the adult limbus or annulus fossae ovalis.

Developmental arrest during septal formation accounts for the

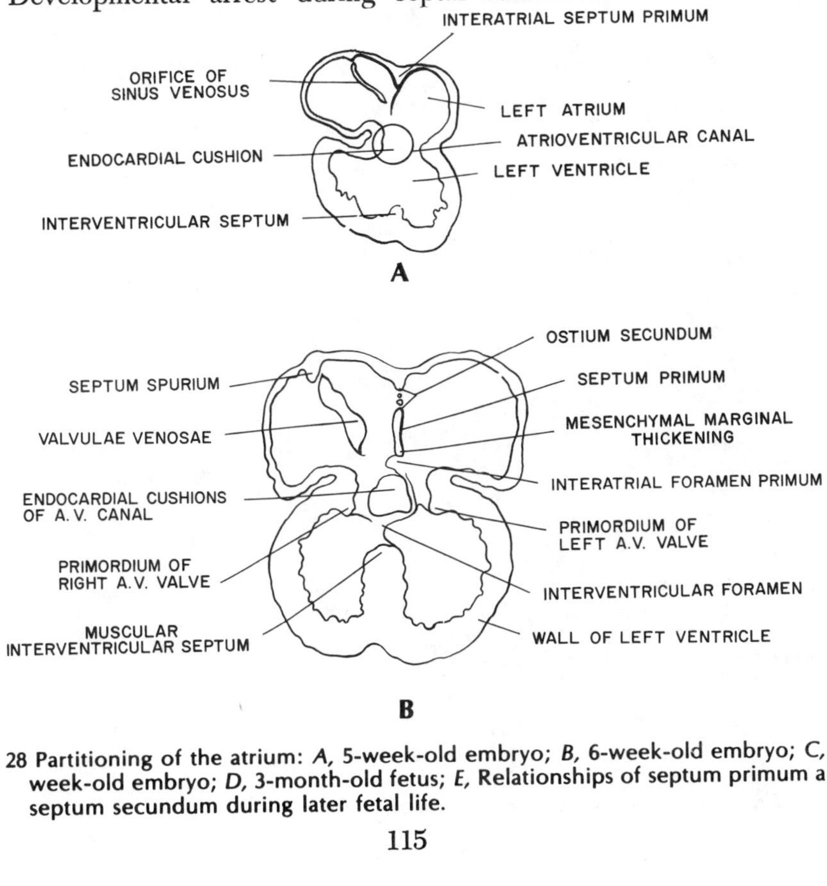

Fig. 28 Partitioning of the atrium: *A*, 5-week-old embryo; *B*, 6-week-old embryo; *C*, 7-week-old embryo; *D*, 3-month-old fetus; *E*, Relationships of septum primum and septum secundum during later fetal life.

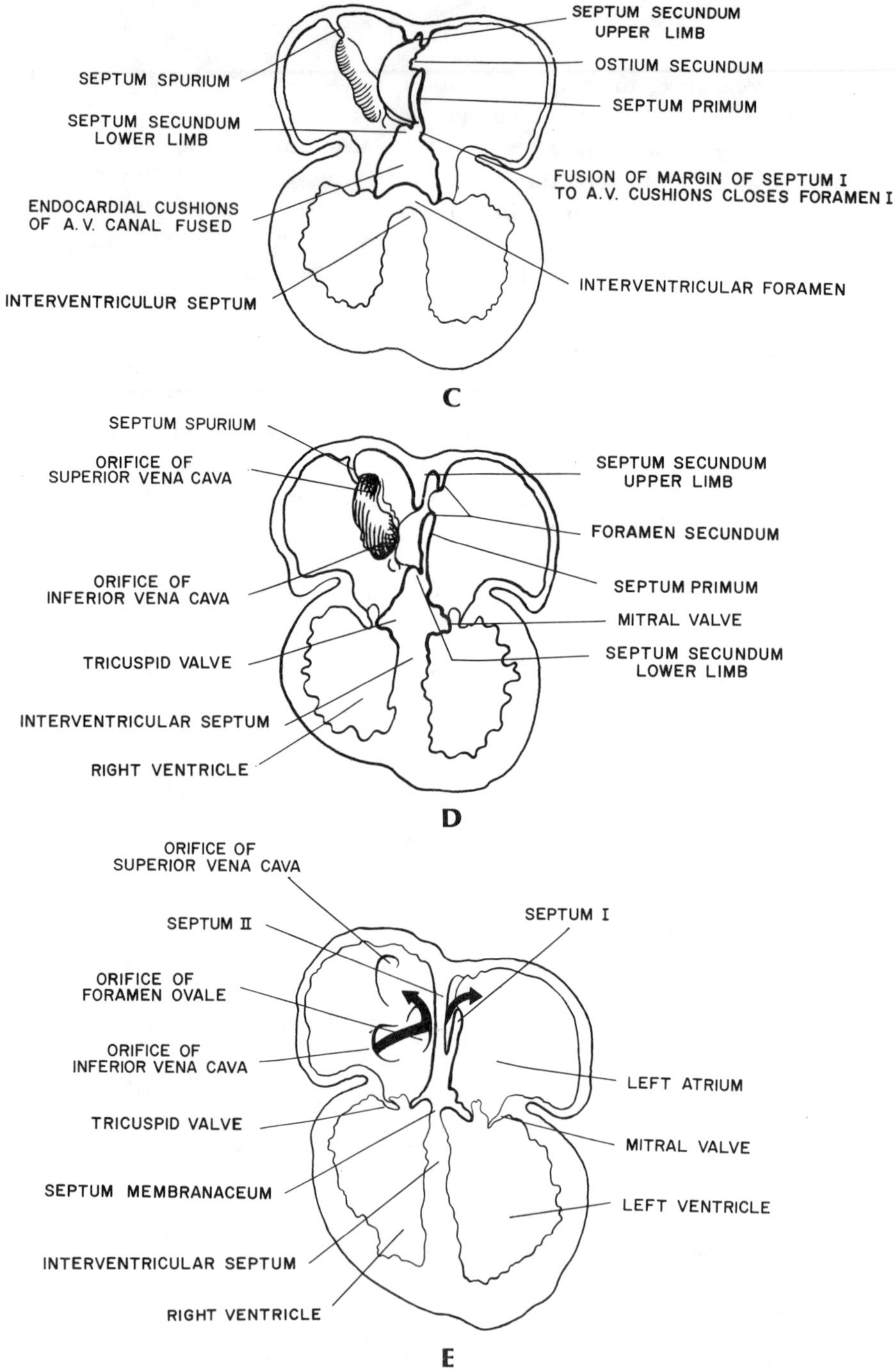

Fig. 28 Continued. (See p. 115 for legend.)

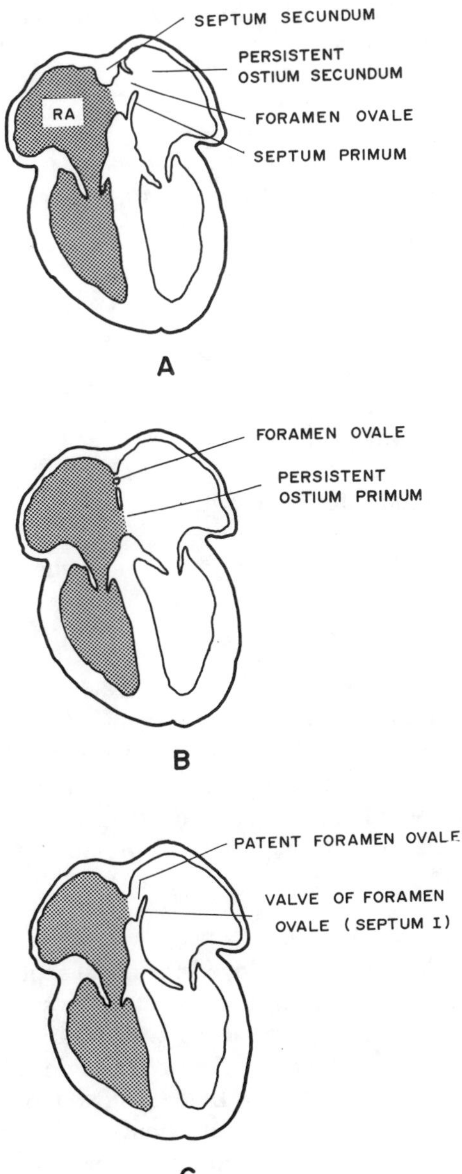

Fig. 29 Classification of interatrial septal defects: *A,* Persistent ostium secundum; *B,* Persistent ostium primum; C, Patent foramen ovale.

variety of atrial septal defects. In general, interatrial septal defects involving the foramen ovale or ostium primum are compatible with life provided they do not include the atrioventricular partition or the valves. Interatrial septal defects make the individual more vulnerable to a number of secondary disturbances.

Classification

In what way(s) may interatrial septal defects be classified? Interatrial defects are categorized by location (Fig. 29):

1. Patent foramen ovale
   a. Uncomplicated
   b. Complicated
      i. Pulmonary stenosis
      ii. Anomalous pulmonary vein drainage
      iii. Mitral stenosis
2. Persistent ostium primum
   a. Central—often fenestrated
   b. Low—has no inferior margin
3. Persistent ostium secundum
4. Localized single or multiple defects anywhere in the interatrial septum

Differential Diagnosis

What malformations may give symptoms similar to atrial septal defects? Atypical patent ductus arteriosus, pulmonary stenosis, anomalous pulmonary veins, and atrioventricularis communis are frequently confused with atrial septal defects and are diagnosed and differentiated from atrial septal defects using a variety of methods.

Treatment

What types of closure are employed for atrial septal defects? Transfixion of the atria using sutures in the plane of the septum; invagination of the atrial appendages or of the lateral wall of an auricle to plug the defect; opening the heart and closing the defect under direct vision utilizing a pump-oxygenator to divert the blood; and temporary occlusion (3 to 4 minutes) of the venae cavae and opening the atria to repair the defect are methods of closure. In addition, atrial septal defects are repaired using a rubber "well" attached to the right atrium and working directly through a pool of blood.

Associated Malformations

What malformations most commonly accompany atrial septal defects? Partial anomalous pulmonary venous return, pulmonary stenosis, mitral valve stenosis, tricuspid atresia, and transposition of the great vessels are most frequently found.

REFERENCES

1. Bailey, C. P., Downing, D. F., Geckeler, G. D., Likoff, W., Goldberg, H., Scott, J. C., Janton, O., and Redondo-Ramirez, H. P.: Congenital interatrial communications: clinical and surgical considerations with a description of a new surgical technic: atrio-septopexy. Ann. Intern. Med. 37:888, 1952.
2. Blalock, A., and Taussig, H. B.: The surgical treatment of malformations of the heart in which there is pulmonary stenosis or pulmonary atresia. J.A.M.A. 128:189, 1945.
3. Bohnholtzer, E.: Beitrag zur Kenntnis einseitiger Lungenagenesien. Beitr. Path. 101:614, 1938.
4. Brock, R. C.: Pulmonary valvulatomy for the relief of congenital pulmonary stenosis. Report of 3 cases. Brit. Med. J. 1:1121, 1948.
5. Cohn, R.: An experimental method for the closure of interauricular septal defects in dogs. Am. Heart J. 33:453, 1947.
6. Cruveilhier, J.: Traité d'anatomie pathologique générale. J. B. Baillière, Paris, 1852.
7. Dennis, C., Spreng, D. S., Jr., Nelson, G. E., Karlson, K. E., Nelson, R. M., Thomas, J. V., Eder, W. P., and Varco, R. L.: Development of a pump-oxygenator to replace the heart and lungs; an apparatus applicable to human patients, and application to one case. Ann. Surg. 134:709, 1951.
8. von Ecker, A.: Berschreibung einiger Fälle von anomaler Communilation der Vorhöfe und Bemerkungen über anomale Communication der beiden Herzhälften überhaupt, derentstehung und deren Einfluss auf die Circulation. Herder, Freiburg, 1839.
9. Fallot, A.: Contribution à l'anatomie pathologique de la maladie bleue (cyanose cardiaque). Marseille Med. 25:77, 1888.
10. Franklin, K. J.: Ductus venosus (Arantii) and ductus arteriosus (Botalli). Bull. Hist. Med. 9:580, 1941.
11. Gibson, G. A.: Clinical lectures on circulatory affections—Lecture I. Persistence of the arterial duct and its diagnosis. Edinburgh Med. J. 8:1, 1900.

12. Gibson, T.: The Anatomy of Humane Bodies Epitomized. Wherein all the parts of Man's Body, with their actions and uses, are succinctly described, according to the newest Doctrine of the most Accurate and Learned Modern Anatomists, ed. 5. Avonshaw and Churchill, London, 1697.
13. Graybiel, A., Strieder, J. W., and Boyer, N. H.: An attempt to obliterate the patent ductus arteriosus in a patient with subacute bacterial endarteritis. Am. Heart J. 15:621, 1938.
14. Gross, R. E., and Hubbard, J. P.: Surgical ligation of a patent ductus arteriosus. Report of first successful case. J.A.M.A. 112:729, 1939.
15. Haight, C., and Towsley, H. A.: Congenital atresia of the esophagus with tracheo-esophageal fistula: Extrapleural ligation of fistula and end-to-end anastomosis of esophageal segments. Surg. Gynec. Obstet. 76:672, 1943.
16. Hoffman, W.: Atresia oesophagi congenita et communicatio inter oesophagum et tracheum. Inaugular dissertation. Julius Abel, Greifswald, 1899.
17. Holder, T. M., McDonald, V. G., Jr., and Wooley, M. M.: The premature or critically ill patients with esophageal atresia. Increased success with a staged approach. J. Thorac. Cardiov. Surg. 44:344, 1962.
18. Keith, J. D., Rowe, R. D., and Vlad, P.: Heart Disease in Infancy and Childhood. Macmillan Company, New York, 1958.
19. Kirklin, J. W., DuShane, J. W., Patrick, R. T., Donald, D. E., Hetzel, P. S., Harshbarger, H. G., and Wood, E. H.: Intracardiac surgery with the aid of a mechanical pump-oxygenator system (Gibbon type): Report of eight cases. Mayo Clin. Proc. 30:201, 1955.
20. Ladd, W. E.: The surgical treatment of esophageal atresia and tracheo-esophageal fistulas. N. Eng. J. Med. 230:625, 1944.
21. Lanman, T. H.: Congenital atresia of the esophagus: A study of 32 cases. Arch. Surg. 41:1060, 1940.
22. Leven, N. L.: Congenital atresia of the oesophagus with tracheo-oesophageal fistula. Report of successful retropleural ligation of fistulous communication and cervical oesophagostomy. J. Thorac. Cardiov. Surg. 10:648, 1940.
23. Lillehei, C. W., Cohen, M., Warden, H. E., and Varco, R. L.: The direct-vision intracardiac correction of congenital anomalies by controlled cross circulation. Results in thirty-two patients with ventricular septal defects, tetraology of Fallot and atrioventricularis communis defects. Surgery 38:11, 1955.

24. MacKenzie, M.: A Manual of Diseases of the Throat and Nose. Wood, New York, 1884.
25. Marioni, P.: Agenesis of the lung. Review and report of three cases. Chest 41:232, 1962.
26. Martin, M.: Report en Exposé des travaux de la societé royale de medicine de Marseilles. 1821.
27. Minetto, E., Galli, E., and Boglione, G.: Agenesia, aplasia, ipoplasia polmonare; rilievi clinici, fisiopathologici e statistici. Minerva Med. 49:4635, 1958.
28. Munro, J. C.: Surgery of the vascular system. I. Ligation of the ductus arteriosus. Ann. Surg. 46:335, 1907.
29. Murray, G.: Closure of defects in cardiac septa. Ann. Surg. 128-843, 1948.
30. Mustard, W. T., and Chute, A. L.: Experimental intracardiac surgery with extracorporeal circulation. Surgery 30:684, 1951.
31. Mustard, W. T., Ravitch, M. M., Snyder, W. H., Jr., Welch, K. J., and Benson, C. D. (eds.): Pediatric Surgery. Yearbook Medical Publishers, Chicago, 1969.
32. Potts, W. J., Gibson, S., Smith, S., and Riker, W. L.: Diagnosis and surgical treatment of patent ductus arteriosus. Arch. Surg. 58:612, 1949.
33. Potts, W. J., Smith, S., and Gibson, S.: Anastomosis of the aorta to a pulmonary artery. Certain types in congenital heart disease. J.A.M.A. 132:627, 1946.
34. Richter, H. M.: Congenital atresia of the oesophagus: An operation designed for its cure. Surg. Gynec. Obstet. 17:397, 1913.
35. Roberts, K. D.: Congenital oesophageal atresia and T.O.F. A review of 36 cases. Thorax 13:116, 1958.
36. Rubin, A.: Handbook of Congenital Malformations. W. B. Saunders Company, Philadelphia, 1967.
37. Schwalbe, E.: Die Morphologie der Missbildungen des Menschen und der Tiere. Gustan Fischer, Jena, 1913.
38. Shaw, R.: Surgical correction of congenital atresia of the esophagus with tracheo-esophageal fistula. J. Thorac. Cardiov. Surg. 9:213, 1939.
39. Stensen, N.: Embryo monstro affinis Parisii dissectus. In Bartholin, T.: Acta Medica et philosofica Hafniencia 1:202, 1671-1672.
40. Stokes, H. L., and Brown, C. J. O.: Agenesis of the lung. Report of a case. Med. J. Aust. 1:49, 1940.
41. Stokes, T. H., and Gibbon, J. H., Jr.: Experimental maintenance of life by a mechanical heart and lung during occlusion of the

venae cavae followed by survival. Surg. Gynec. Obstet. 91:138, 1950.

42. Vogt, E. C.: Congenital esophageal atresia. Amer. J. Roentgen. 22:463, 1929.

43. Welsch, K.: Ein Fall von Mangel der rechten Lunge. Frankfurt Z. Path. 36:192, 1928.

44. Wright, J. S., and Johnston, J. B.: Cardiac surgery in infants. Med. J. Aust. 2:80, 1968.

Supplement: Cardiac Catheterization

Cardiac catheterization is the passage of a small-caliber tube (the catheter) into a vein or artery, normally in the arm or thigh, and through the blood vessels into the heart in order to obtain blood samples, determine intracardiac pressure and cardiac output, and detect and/or confirm cardiac defects. Fritz Bleichroeder in 1905 was the first to pass catheters without the use of X-ray into his own veins and into the veins and arteries of dogs.

Defects in the heart investigated using cardiac catheterization are of four types: (1) obstruction in valves, atria, ventricles, or great vessels; (2) incompetence of valves; (3) septal defects and aortopulmonary communications; and (4) abnormal connections of veins, great arteries, or heart chambers.

Techniques of cardiac catheterization for the study of congenital heart disease in infants differ only in detail from those employed in older children and adults. However, the small size of the infant and the fact that he is subjected to such tests as cardiac catheterization only if he is sufficiently ill to require early surgery account for differences in technique. The size of the infant restricts the availability of veins and arteries large enough to accept a cardiac catheter and his small blood volume requires careful hemostasis of blood gas analysis on samples which should not exceed 1 ml. The number of diagnostic tests and time spent on each should be rigidly controlled to avoid overexertion of the infant.

The following procedure for cardiac catheterization (Fig. 30) is that used on the infant or young child:

1. Children under twelve years of age are always electrocardiogram

CARDIAC CATHETERIZATION

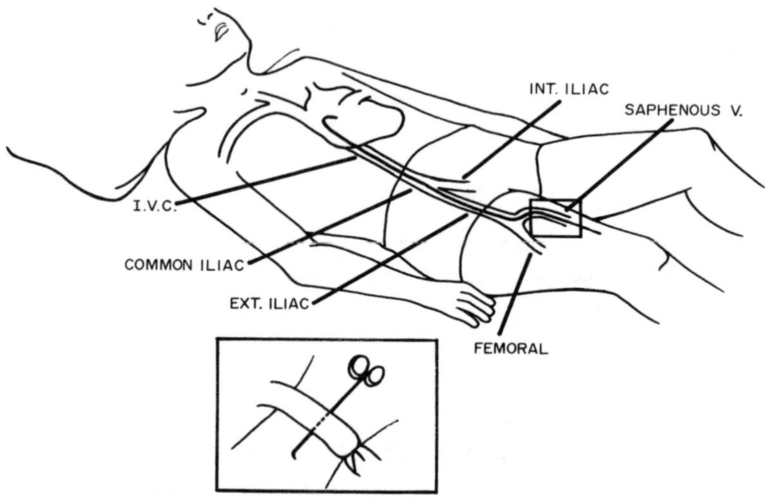

Fig. 30

(ECG) monitored and sedated because usually it is not possible to distract them for a sufficient length of time to complete the procedure.

2. Selection of vessel for passage of catheter:
   a) right heart catheterization using the right saphenous vein is preferred; alternate veins include the right basilic (along the medial border of the biceps), antecubital, axillary, or the external jugular vein as a last resort.
   b) retrograde left heart catheterization usually employs the right brachial, common carotid, subclavian, axillary, ulnar, radial, or femoral artery.
3. Incision is made in a skin fold (Langer's lines) overlying the selected vessel.
4. Isolation of vessel:
   a) a ligature is placed at each end of the vessel; the distal ligature is tied and remains intact during the procedure while the proximal ligature is secured without tying (e.g., using a hemostat).
   b) the vessel thus isolated is brought to the surface of the skin using a curved hemostat to dissect it from the surrounding subcutaneous tissue.

5. Incision of vessel. The vessel must fit snugly around the catheter to avoid continuous retrograde bleeding.
6. Insertion of catheter (*should never be forced*). A No. 5 catheter is usually employed, or in infants weighing less than 10 pounds a No. 4 may be used. As the catheter approaches the heart it is followed by direct x-ray vision utilizing a fluoroscope and by an image intensifier to relay the x-ray to TV sets.

Right Heart Catheterization

The catheter is introduced into the right saphenous vein and continues through the femoral vein → external iliac vein → common iliac vein → inferior vena cava → right atrium → right ventricle → pulmonary artery. Then the right heart is explored to locate a defect.

Retrograde Left Heart Catheterization

There are two possible pathways:

1. The catheter is introduced into the right brachial artery and continues through the junction of the right subclavian and common carotid arteries → brachiocephalic artery → ascending aorta (extreme care not to enter coronary arteries) → left ventricle → left atrium.
2. The catheter is introduced into the left femoral artery and continues through the external iliac artery → common iliac artery → descending aorta → arch of aorta → ascending aorta → left ventricle → left atrium.

Once the catheter enters the left atrium, the left heart is explored to locate a defect.

Remove the catheter. The vessel is stitched and the skin sutured. An angiocardiogram (roentgenography of the heart and great vessels following intravenous injection of an opaque dye) usually accompanies catheterization to facilitate demonstration of the defect. Before undertaking cardiac catheterization in the young child, it is essential to have as nearly complete and accurate a diagnosis as possible.

ABDOMEN

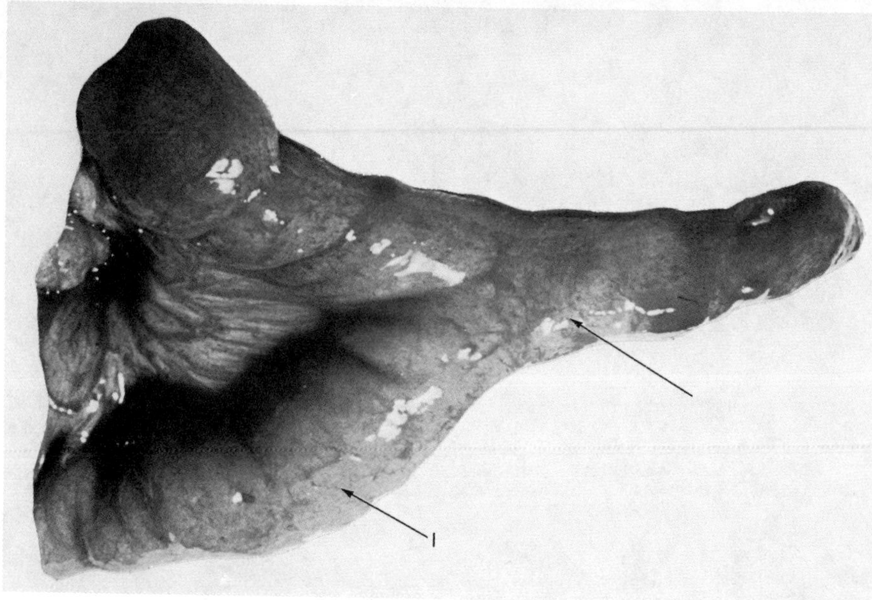

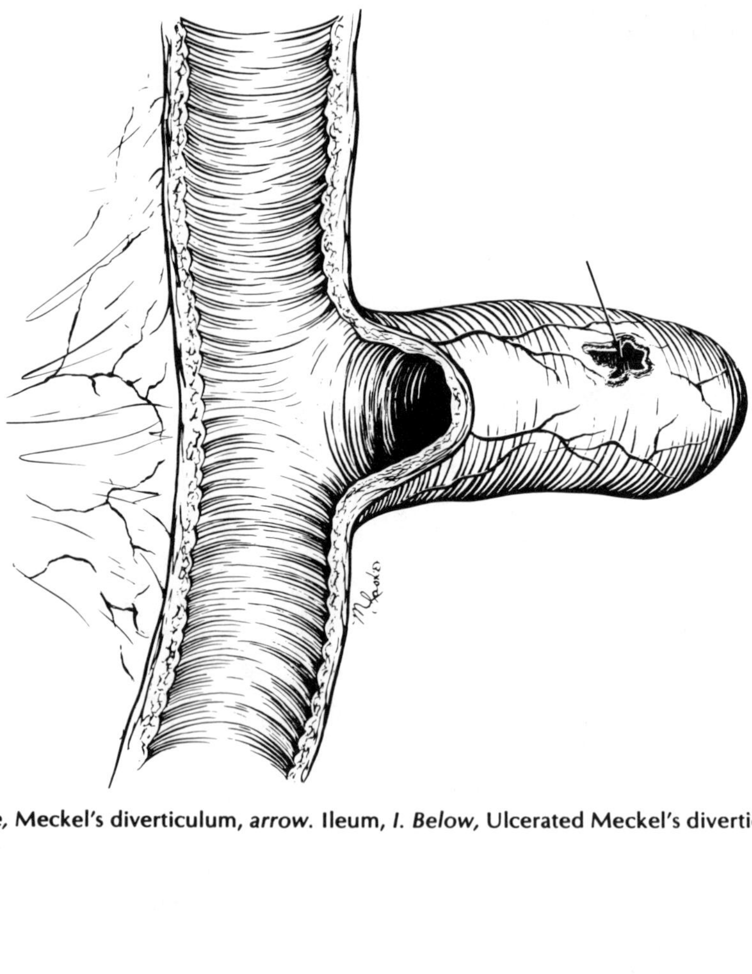

Above, Meckel's diverticulum, *arrow*. Ileum, *I*. *Below,* Ulcerated Meckel's diverticulum, *arrow*.

16

Meckel's Diverticulum

A 5-year-old boy is brought to a physician's office with a history of epigastric and lower abdominal pain. The mother states that the attacks have occurred once a month for six months and lasted one to two days. The last attack was accompanied by vomiting and passage of a bloody stool. The patient is admitted to Children's Hospital.

EXAMINATION

After 12 hours the pain and maximal tenderness do not shift to the right lower quadrant, but remain midline. Roentgenologic examination using barium enema and meal reveals nothing. With continued rectal bleeding, an exploratory laparotomy is considered necessary.

TREATMENT

When the abdomen is opened, the appendix is found retrocecal and considered to be normal. Further exploration along the gastrointestinal tract reveals a perforated diverticulum 3 cm. long and approximately 30 cm. from the ileocecal valve on the antimesenteric border of the ileum. Removal of the ulcerated diverticulum and resection of a short segment of ileum are performed. The peritoneal cavity shows no signs of inflammation and the postoperative course is uneventful. Microscopic examination of the diverticulum reveals patches of ulcerated gastric epithelium associated with the normal ileal epithelial lining.

DIAGNOSIS

Meckel's diverticulum with ulcerated heterotopic gastric epithelium.

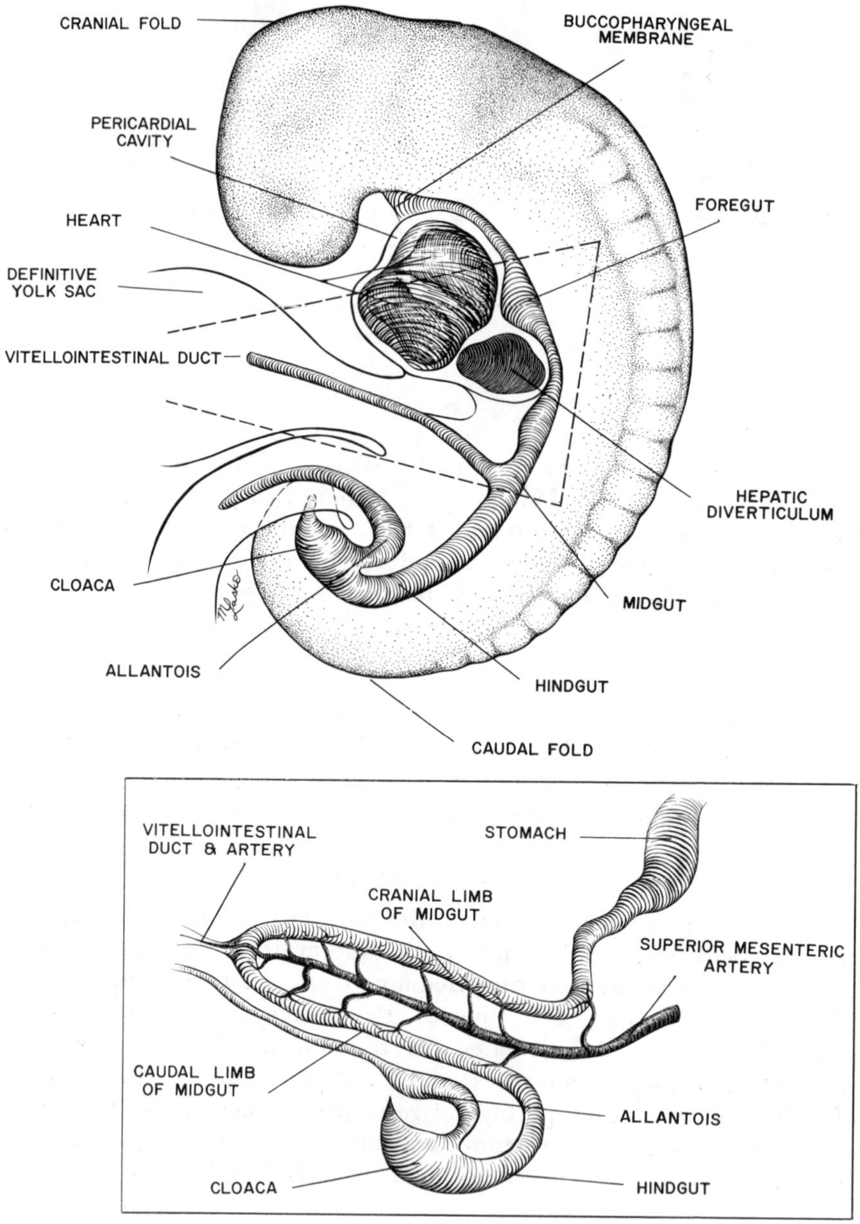

Fig. 31 Developing gastrointestinal system in 4-week-old embryo. *Inset:* Vitellointestinal duct relationships in 5-week-old embryo.

DISCUSSION

The congenital malformation of the intestine described by J. F. Meckel[19] in 1808, Meckel's diverticulum, occurs in approximately 2 per cent of the population.[25]

Embryology

How do you explain embryologically the presence of this diverticulum? During the fourth week of development (Fig. 31), the midgut is in free communication with the definitive yolk sac through the vitellointestinal duct (omphaloenteric, omphalomesenteric). Growth of the midgut during the fifth week of development (Fig. 31 inset) forms a simple U-shaped loop with the narrowed vitellointestinal duct entering at the loop's apex and dividing the midgut into a cranial and a caudal portion. The cranial portion gives rise to the duodenum caudal to the entrance of the common bile duct, the jejunum, and the first two thirds of the ileum, while the caudal portion gives rise to the remainder of

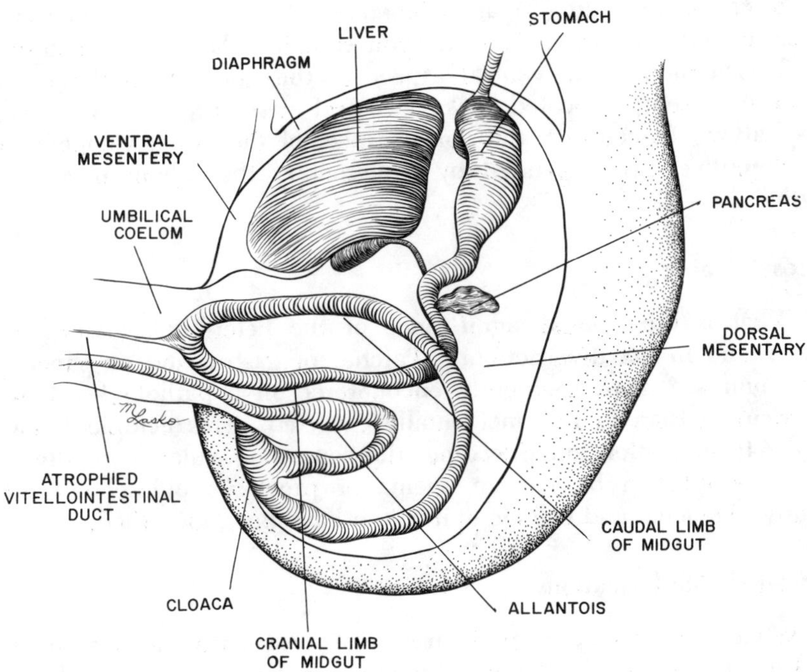

Fig. 32 Developing gastrointestinal system in 7-week-old embryo.

the ileum, cecum, appendix, ascending colon, and the proximal two thirds of the transverse colon. Herniation of the midgut into the umbilical coelom, with accompanying elongation and partial rotation, occurs during the sixth to seventh weeks (Fig. 32). While the midgut lies in the umbilical coelom, the vitellointestinal duct normally atrophies and disappears. The continuation of the intra-abdominal portion of the vitellointestinal duct, due to arrested development and lack of resorption, once the midgut has returned to the abdominal cavity in the tenth week, constitutes Meckel's diverticulum. Although Meckel's diverticulum is the most common malformation of the vitellointestinal duct, failure of obliteration of this duct may also result in: (1) fistulas; (2) enterocysts; (3) fibrous bands extending from the small intestine to the umbilicus; (4) mucosal polyps; and (5) sinuses.[29]

Appearance

What characteristics verify the presence of Meckel's diverticulum? Three distinguishing features of the diverticulum are: (1) the same structural composition (mucosa, submucosa, muscularis, serosa) as the wall of the small bowel; (2) an antimesenteric position on the distal part of the ileum; and (3) an independent blood supply from the superior mesenteric artery. How do you explain a lack of mesentery on the diverticulum? The ventral aspect of the gut, where the diverticulum, if present, is located, lacks a ventral mesentery even as a transitory feature. In contrast, the dorsal aspect of the gut developmentally and definitively is suspended by a mesentery throughout most of its length.

Mucosal Lining

What is the clinical significance of the heterotopic tissue sometimes found in the diverticulum? Patches of gastric and/or pancreatic epithelium are more frequently encountered in a pathologic Meckel's diverticulum than in one incidentally detected. This ectopic lining is subject to ulceration and bleeding, the source of which normally cannot be established by barium enema or proctosigmoidoscopy. Infrequently, duodenal and colonic epithelium have been identified.

Associated Malformations

What malformations[8, 21, 29] are most frequently associated with Meckel's diverticulum? The most common malformations are related to the digestive system: (1) gut malrotation; (2) umbilical hernia; (3) omphalocele; (4) volvulus; (5) esophageal atresia; and (6) cleft palate. Hereditary factors resulting in a Meckel's diverticulum are unknown at present.

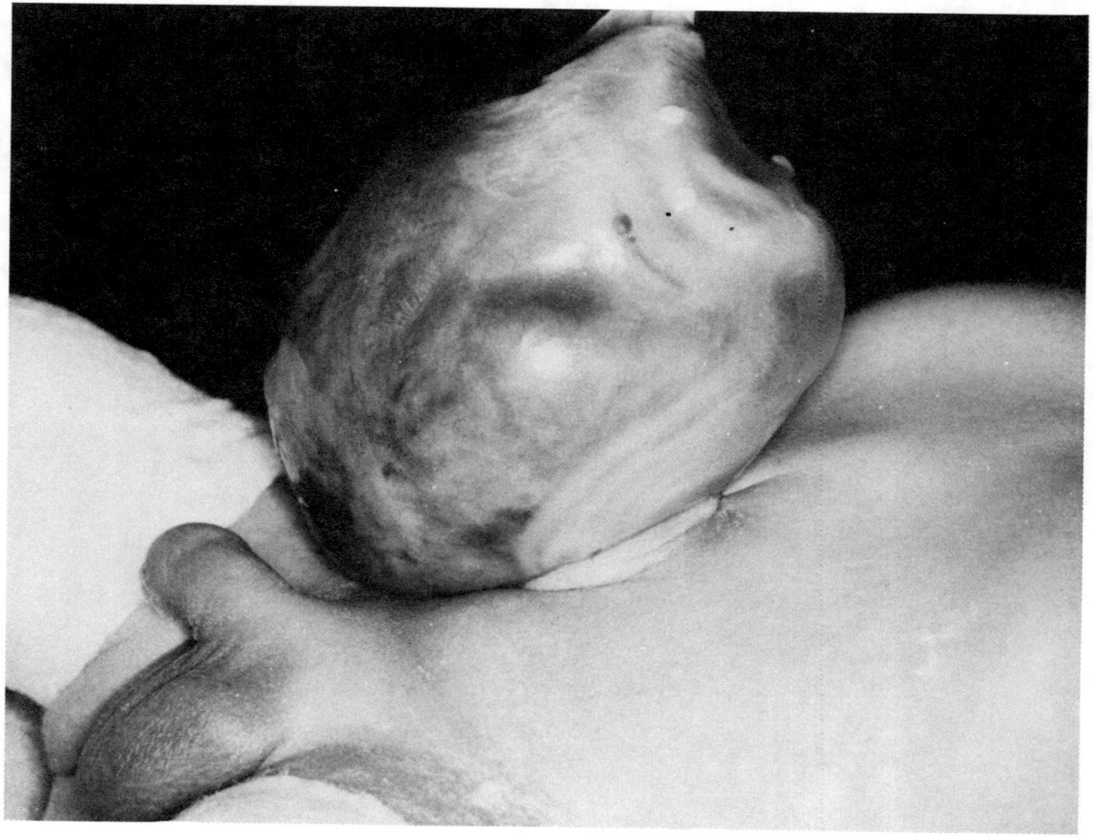

Omphalocele.

17

Omphalocele

A premature male infant (2,500 gm.), the product of an uncomplicated pregnancy and delivery of a 22-year-old gravida I, para I

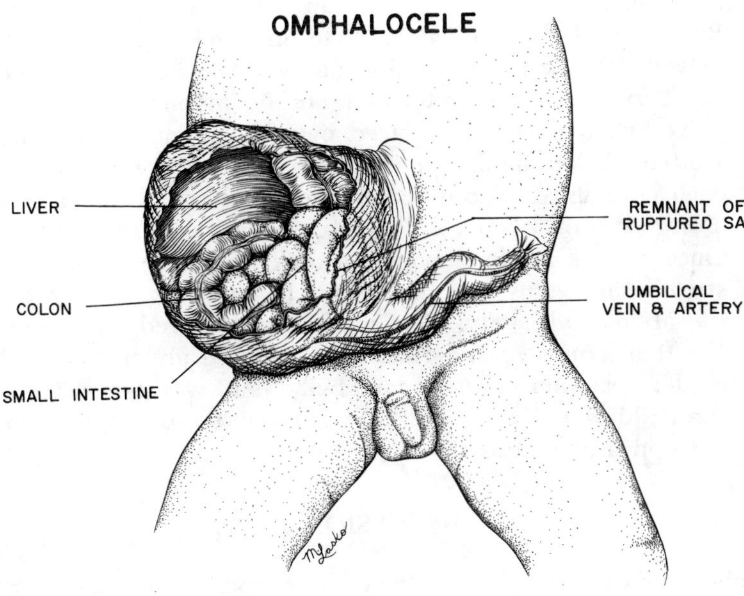

Fig. 33

mother, was born with an omphalocele (Fig. 33). The omphalocele sac ruptured during delivery, exposing the herniated contents of the abdominal cavity. A sterile wet-dressing was immediately applied to the contents. A thin rubber catheter was passed through the nose to the stomach, and the stomach was emptied every 15 minutes to prevent vomiting, abdominal distention, and a consequent increase in the size of the omphalocele. The infant was taken directly to surgery.

EXAMINATION

Examination revealed a shallow abdominal cavity and an 8-cm. defect in the abdominal wall through which portions of large and small bowel and liver protruded. There was evidence of obstruction, malrotation, and lack of fixation of the gut. Remnants of the omphalocele sac disclosed a thin, translucent, avascular membrane; the umbilical cord was inserted at the apex of the sac, while the umbilical vein and arteries were within its walls. No additional malformations were noted.

TREATMENT AND FURTHER COURSE

Surgical correction was undertaken in two stages due to the large size of the defect.

1. Remnants of the omphalocele sac and umbilical cord were removed. The skin of the abdominal wall was insufficient to cover the herniated mass and could not be mobilized; therefore, a polyethylene sheet was placed over the herniated contents. A knitted Teflon mesh applied over the sheet was attached to each laterally displaced rectus abdominis muscle, thus creating a large ventral hernia. Initial surgery was well tolerated.
2. Once a week for four months a tuck was taken in the mesh to stretch the rectus abdominis muscle and to gradually enlarge the abdominal cavity. The mesh was removed by the end of the fourth month, and final repair of the abdominal wall, including excision of redundant belly skin, was completed when the child was 2 years old. Further intestinal complication had not appeared at this time.

DISCUSSION

Ambroise Paré (1634)[17] gave the earliest report of this malformation, which occurs in 1 of 6,600 births.[25] The first successful surgical treatment of an omphalocele was reported by Hey in 1803.[13] In 1887 Olshausen[23] recommended a two-stage repair of large omphaloceles, a technique that was revived and modified by Gross (1948).[15] Conservative therapy using alcohol dressings, introduced by Ahlfeld in 1899[1] was renewed by Grob in 1963.[13] The more recent methods of repair utilize plastic mesh and split-skin grafts for closure.[20, 26]

Embryology

How do you explain developmentally the presence of an omphalocele (exomphalos, umbilical eventration, funicular hernia of the umbilicus, congenital amniotic hernia)? Two developmental processes contribute to this defect. First, the 6- to 12-week-old embryo's intraembryonic coelomic cavity extends to the base of the umbilical cord and for a short time contains the greater part of the midgut while the abdominal cavity is enlarging. Coincident with this enlargement, the midgut normally undergoes rotation and returns to the expanded abdominal cavity during the tenth week of development, whereupon portions of it become fixed to the posterior body wall. Secondly, the closing of the body wall as the midgut is drawn into the abdominal cavity is accomplished by the circumferential folding of the (1) cephalic, (2) caudal, and (3) two

UMBILICAL HERNIA

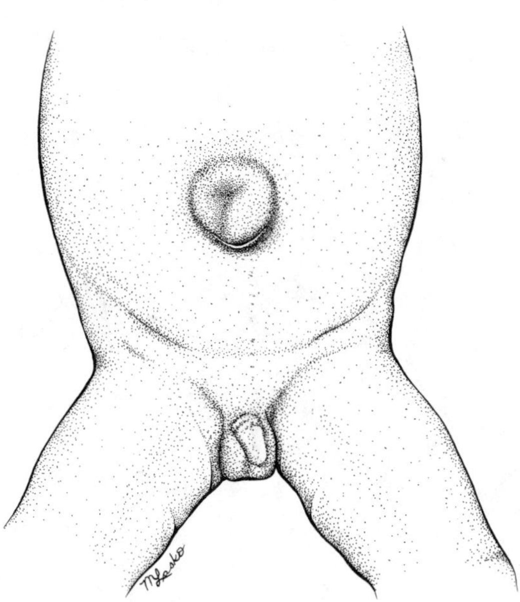

Fig. 34

lateral body folds. The cephalic fold forms the thoracic and epigastric walls, while the caudal and lateral folds form the hypogastric and lateral body walls, respectively. The apex of these folds forms the future umbilical ring. Arrested development and lack of fusion of the lateral folds results in a circumscribed defect with various amounts of herniated intestine protruding. The defect is covered by a sac composed of the abdominal peritoneal extension and amnion, with a layer of Wharton's jelly interposed between them. The umbilical cord lies at the apex of this sac, the latter being devoid of a skin covering.

What abdominal wall defects are sometimes identified as an omphalocele?

1. Umbilical hernia (Fig. 34)—a defect of small to moderate size in the rectus muscles and fascia through which the peritoneal sac, covered only by skin, protrudes.
2. Gastroschisis (Fig. 35)—a failure of a portion of the somatic layer in the lateral body fold to differentiate, resulting in a normal umbilical ring but an adjacent defect, usually on the right side.

GASTROSCHISIS

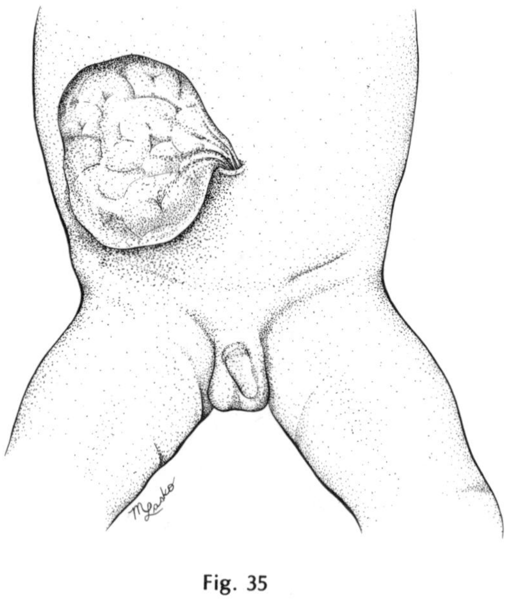

Fig. 35

Classification

How are omphaloceles classified? Classification[24] is based upon: (1) the size of the abdominal wall defect: (a) small to medium < 5 cm., (b) large > 5 cm.; (2) associated malformations requiring surgical correction; and (3) the condition of the omphalocele sac: (a) ruptured—pre- or postnatal, (b) unruptured.

Omphalocele Repair

What treatments are utilized for repair of an omphalocele? Treatment[20, 24, 35] is based upon the extent of the defect. A *primary closure* is used for a small defect < 5 cm. where the amount of protruding viscera is relatively slight. The viscera are replaced, the sac excised, and the final closure is made, taking care not to increase the intra-abdominal pressure and hence interfere with respiration.

A *two-stage repair* is used for a large defect > 5 cm. If an unruptured sac exists, the umbilical cord is amputated flush with the sac, surrounding skin is excised and loosened except over the chest, and the incision is closed, thus creating a large ventral hernia. The second stage is repair of the hernia when the child is two to three years old and the abdominal cavity can accommodate the herniated contents. Disadvantages of this method are: (1) the inability to inspect the abdominal viscera for malformations which could cause complications at a later time; (2) the occurrence of adhesions between the sac and the intestine; (3) the necessity of multiple operations using general anesthesia; and (4) the possibility of infection resulting in rejection of the Teflon mesh.

If the sac is ruptured, the two-stage method of repair can also be utilized, as exemplified by the present case.

Conservative treatment consists in painting the omphalocele sac with an antiseptic that promotes formation of an eschar, thus protecting against infection. Epithelial proliferation and contracture close the abdominal wall defect (10-cm. defect closed in approximately four months). This method is preferable for defects < 5 cm. and for premature infants and infants requiring surgical correction of extra-abdominal defects. The disadvantages of conservative treatment are the possible cracking of the eschar, infection, and subsequent peritonitis, while the important advantage is the slow reduction of the omphalocele.

Replacement of Viscera

What three factors must be considered in reducing the omphalocele by replacement of the abdominal viscera? These factors are: (1) elevation of pressure on both sides of the diaphragm with a decrease in vital capacity and subsequent pulmonary complications; (2) pressure on the inferior vena cava and the mesenteric and portal systems, causing circulatory failure; and (3) temporary intestinal obstruction.

Associated Malformations

What malformations are commonly associated with an omphalocele? This defect characteristically has malrotation and lack of fixation of the midgut. Additional abnormalities include Meckel's diverticulum, intestinal atresia, urinary tract or genital malformations, hydrocephalus, anencephaly, clubfoot, and cleft lip.

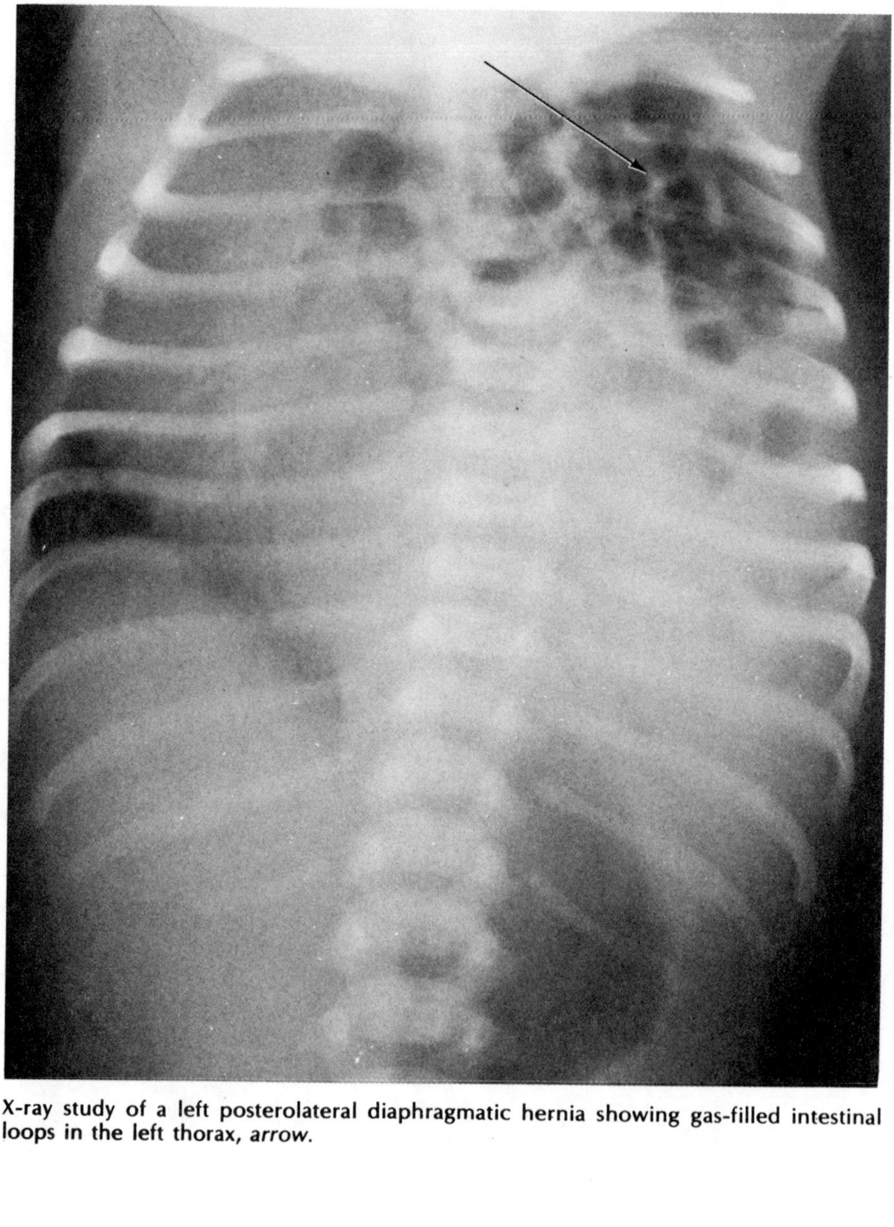

X-ray study of a left posterolateral diaphragmatic hernia showing gas-filled intestinal loops in the left thorax, *arrow*.

18

Congenital Diaphragmatic Hernia

A 22-hour-old male infant born at term (3,200 gm.) of an uncomplicated pregnancy was transferred to Children's Hospital because of respiratory difficulty (dyspnea).

EXAMINATION

Examination showed the infant to possess subcostal and sternal retractions, scaphoid abdomen with a lack of bowel sounds, and a slight cyanosis that was relieved by oxygen. X-rays revealed gas-filled intestinal loops in the left thorax accompanied by a mediastinal shift to the right, and hypoplasia of the left lung.

DIAGNOSIS

Congenital left posterolateral diaphragmatic hernia through the foramen of Bochdalek.

TREATMENT

Nasogastric suction was applied prior to surgery to help decompress the intestines. An abdominal incision revealed the small and large intestines up to the transverse colon displaced through a defect in the posterior left half of the diaphragm (Fig. 36). Exploration of the abdominal area indicated incomplete intestinal rotation. The abdominal viscera were subsequently removed from the left pleural cavity by gentle pulling and placed in an approximate normal position in the abdomen. No remnant of a hernial sac was noted. The diaphragmatic defect was repaired in two layers (pleural and peritoneal), and the abdominal wall

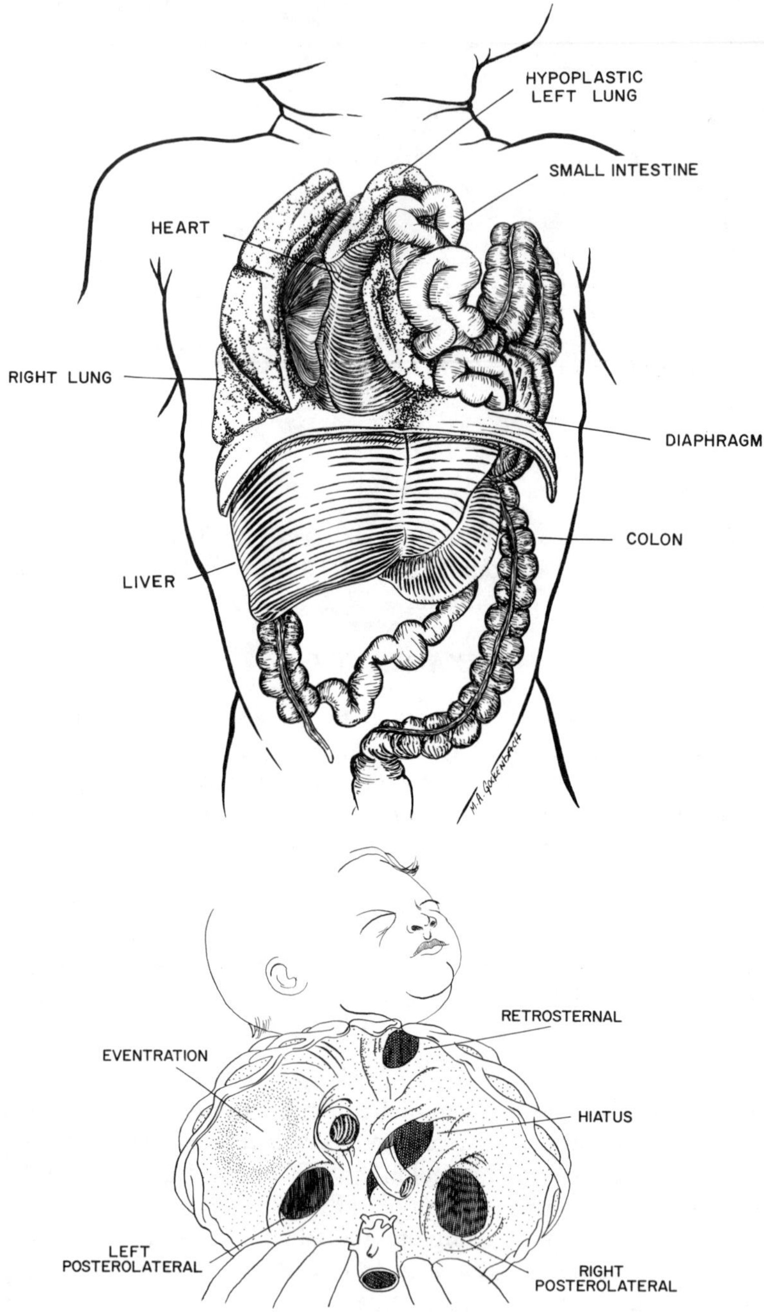

Fig. 36 *Above,* Displacement of abdominal viscera in a left-sided diaphragmatic hernia; *Below,* Commonest diaphragmatic defects through which hernias occur.

was closed with some difficulty. Postoperatively, expansion of the left lung did not occur and the patient expired.

DISCUSSION

Ambroise Paré (1575)[21] gave the first description of a diaphragmatic hernia, which occurred in an adult. The pathology and classification of the various diaphragmatic hernias were described by Sir Astley Cooper (1844).[6] In 1848 Bochdalek[3] clarified the embryology of the type of diaphragmatic defect discussed in the present case. This defect is responsible for 1 in 50 neonatal deaths in the first three days of life and comprises 8 per cent of all major fatal congenital anomalies.[24, 25]

Embryology

Developmentally, how do you explain the present diaphragmatic defect? A hernia through the posterolateral foramen of Bochdalek involves three intimately timed embryological processes: (1) formation of the diaphragm that separates the pleural from the peritoneal cavity; (2) return of the midgut to the abdominal cavity from the umbilical stalk; and (3) lung formation. In the seventh week of development, the midgut undergoes a rapid growth; since the fetal abdomen is two small to contain the entire midgut, the latter passes into the umbilical stalk where it continues the processes of complex rotation around the superior mesenteric artery. By the tenth week the midgut returns to the left side of the abdomen and continues its rotation, with part of the midgut eventually coming to lie on the right side.

The following explanation of the formation of the intraembryonic coelom and the body cavities is essential to an understanding of the present case.

In an early somite embryo (Fig. 37A), small spaces lined with mesothelium appear in the lateral plate mesoderm. These spaces prolongate cranially within the embryo, become confluent, and eventually fuse across the midline in the cardiogenic mesoderm. The resulting horseshoe-shaped closed cavity is the intraembryonic coelom. Later, at the lateral edges of the caudal extremities of the intraembryonic coelom, a communication with the extraembryonic coelom will ensue. The intraembryonic coelom is lined by splanchnopleure and somatopleure (Fig. 37B).

The cranial portion of the intraembryonic coelom crosses the midline of the embryonic disk, transversely and anterior to the buccopharyngeal membrane, to form the pericardial cavity. The cardiogenic plate develops in the splanchnopleuric mesoderm, and the endothelial heart tubes develop from this plate and designate the site of the future pericardial cavity.

With the formation of the head fold, the endothelial heart tubes

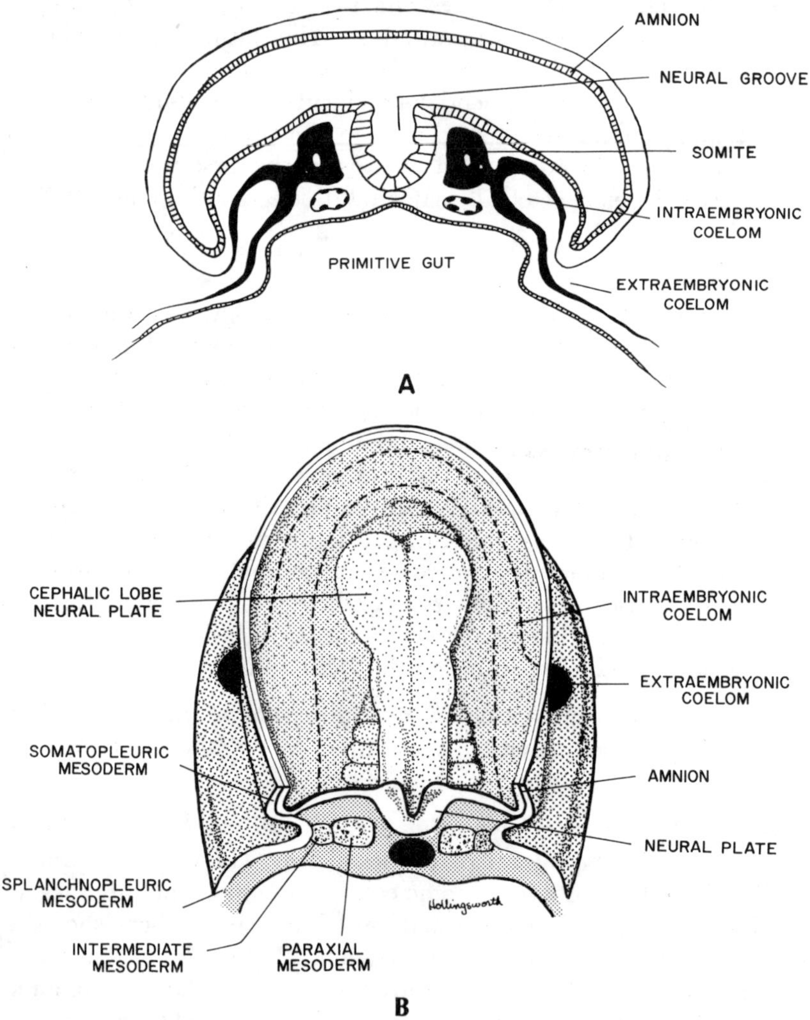

Fig. 37 Division of coelom; *A,* X-s 3-week-old embryo; *B,* Dorsal view, 3-week-old embryo; *C,* X-s 5-week-old embryo; *D,* Sagittal view, 5-week-old embryo showing continuity of cavities; *E,* Sagittal view, 6-week-old embryo; *F,* Sagittal view 7-week-old embryo. (From Hamilton, W. J., Boyd, J. D., and Mossman, H. W.: Human Embryology, ed. 3. W. Heffer & Sons, Cambridge, England, 1964, with permission.) (Illustrations are continued on the three following pages.)

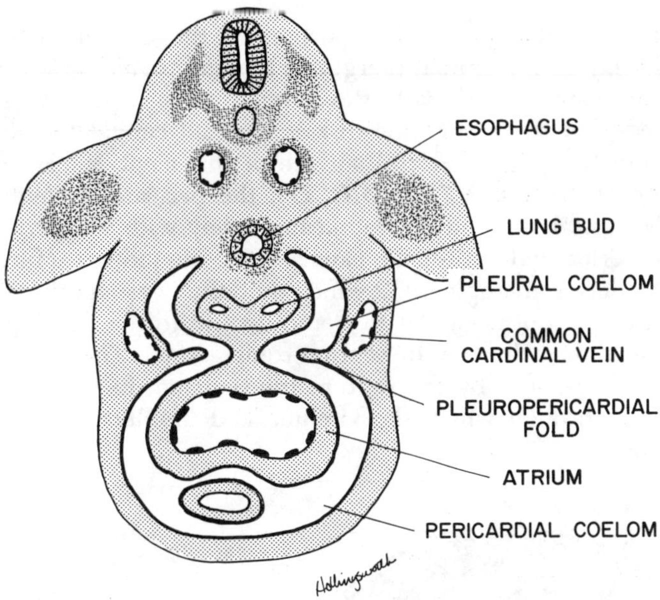

C

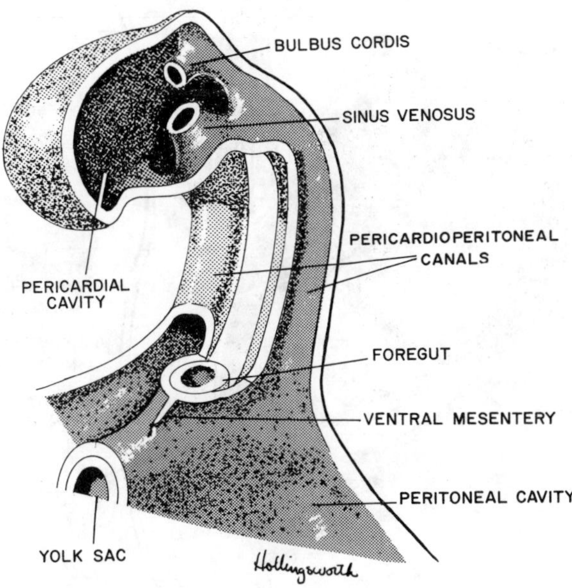

D

Fig. 37 (continued. Illustrations continue on the two following pages. See facing page for legend.)

and the transverse pericardial portion of the coelom, both of which originally lay in the cranial margin of the embryonic disk, are bent ventrally and caudally beneath the foregut. Consequently, the original cranial edge of the embryonic disk is situated caudal to the pericardial cavity, and the original dorsal surface of this cavity faces ventrally. At this stage (fifth week) (Fig. 37C, D), the pericardial cavity is proportionately quite large and communicates freely with the peritoneal cavity via the pericardial-peritoneal canals. The mesoderm (fused splanchnopleure and somatopleure), which initially was cranial to the pericardial cavity, is now caudal to it and ventral to the developing foregut. This mass of mesoderm is the septum transversum. The pericardial cavity is bounded caudally by the septum transversum, in which the liver develops and through which the vitelline and umbilical veins pass to the sinus venosus.

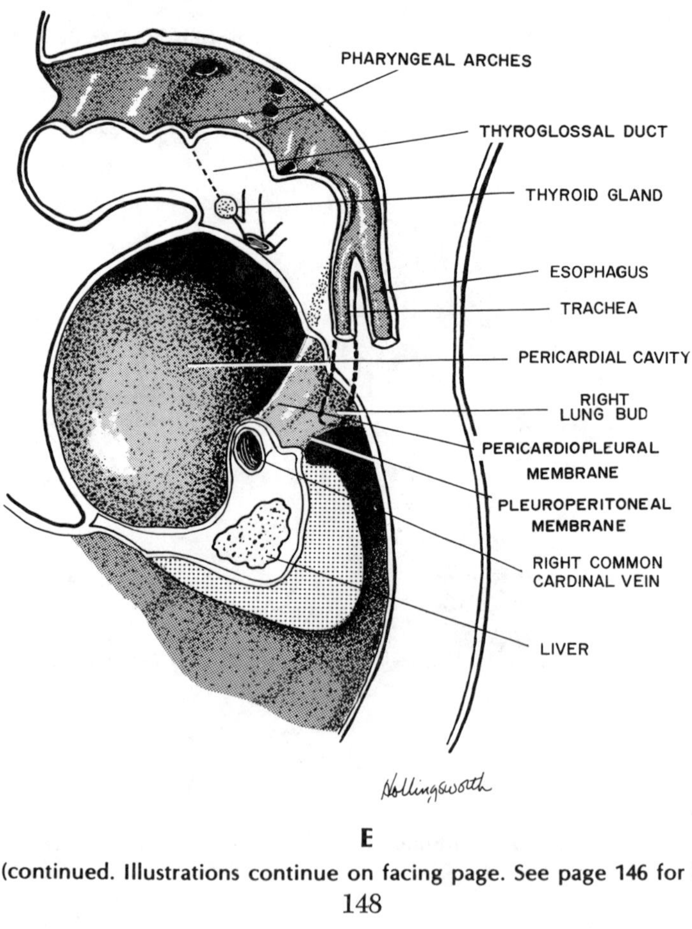

E

Fig. 37 (continued. Illustrations continue on facing page. See page 146 for legend.)

After formation of the head fold, the pericardial cavity remains in communication with the intraembryonic coelom by a constricted canal, the coelomic duct. This duct passes dorsal to the septum transversum. As the foregut lengthens, the coelomic ducts also elongate; each duct is invaginated on its medial aspect by a lung bud and the ducts become the pleural cavities. As the lung buds develop, the heart descends relative to the other structures, so that the common cardinal veins (duct of Cuvier) pass in a ventromedial direction to the sinus venosus (Fig. 37E), forming the pericardiopleural membrane that divides the pericardial from the pleural cavity.

On each side of the embryo, the peritoneal cavity freely communicates with the extraembryonic coelom. The early position of the gut and its mesentery in the midline divides the peritoneal cavity into a right and a left half. These halves intercommunicate ventral to the

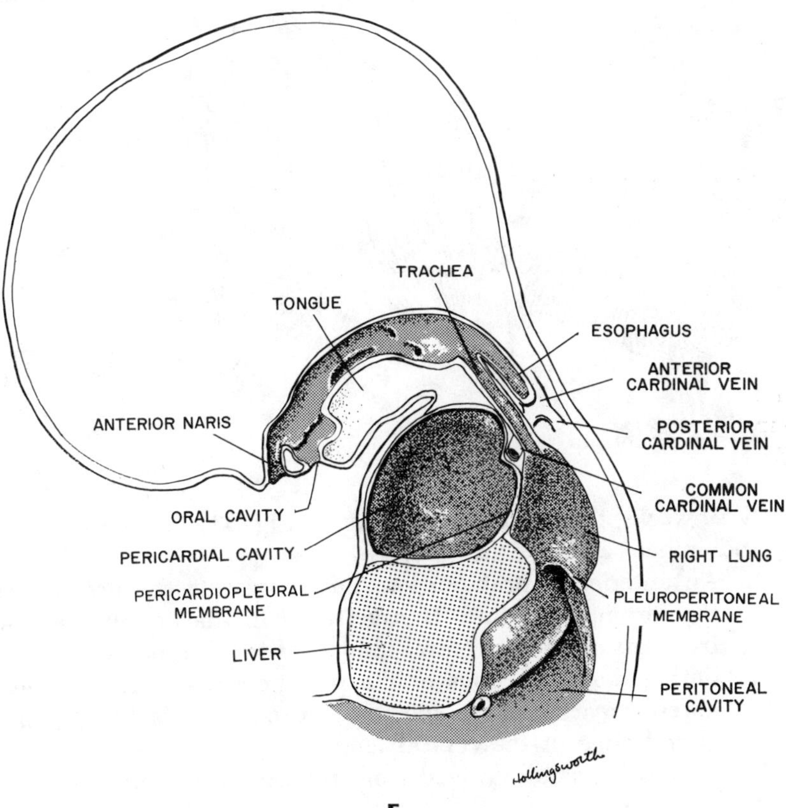

F

Fig. 37 (continued. See page 146 for legend.)

gut (Fig. 37D). With expansion of the peritoneal cavity and narrowing of the umbilicus, the communication between the intra- and extra-embryonic coelom is progressively constricted and finally obliterated as the umbilical cord loses its communication with the peritoneal cavity. Concurrently, as the pericardial cavity is closed off from the intra-embryonic coelom, the mesonephros becomes situated more caudally in the abdominal region. A ridge of mesoderm, the pleuroperitoneal membrane, connects the cranial pole of the mesonephros and the adrenal gland to the edge of the septum transversum. As the pleural cavity enlarges, the pleuroperitoneal membrane becomes extensive and eventually will close off the pleural from the peritoneal cavity (Fig. 37F).

After closure of the pleuroperitoneal communication at approxi-

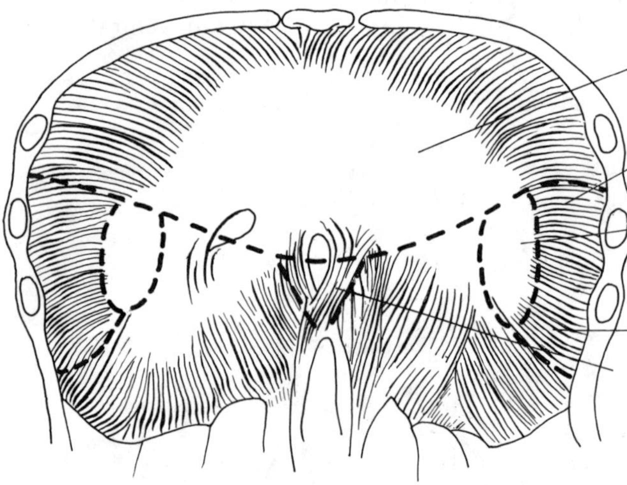

Fig. 38 Adult diaphragm.

mately 10 weeks, the resulting diaphragm (Fig. 38) between the thoracic and abdominal cavities consists of:

1. An anterior central portion that represents the greater part of the original septum transversum; in the adult the septum transversum is the central tendon of the diaphragm.
2. Paired dorsal portions of the pleuroperitoneal membrane. The anterior margin of this membrane is continuous with the postero-lateral edge of the central tendon.
3. A dorsal portion derived from the esophageal mesentery.
4. An extension of the lining and muscular tissue from the body wall.

Premature return of the midgut to the abdominal cavity or delayed closure of the diaphragm may produce a diaphragmatic hernia. The inter-relationship of the developing diaphragm and the gastrointestinal tract determines the type of diaphragmatic hernia and explains the frequent association of diaphragmatic hernias with malrotation of the gut. Since both bronchial and alveolar development of the lung lags behind development of the midgut, the severity of pulmonary hypoplasia (defective or incomplete development) is determined by whether the midgut passes into the thorax at the eighth (severe) or twelfth (mild) week of development.

Classification

A diaphragmatic hernia through the foramen of Bochdalek constitutes a surgical emergency in the newborn. What other diaphragmatic defects, congenital or acquired, are recognized? Hernias of the diaphragm have been categorized;[7,14] the classifications differ slightly, but most investigators are in general agreement on the following types (Fig. 36B): (1) retrosternal or retrocostosternal hernias through the foramen of Morgagni; (2) posterolateral defects; (3) eventration; (4) esophageal hiatus hernia; (5) extensive defects involving most or all of the hemidiaphragm; and (6) herniation into the pericardium.

Surgical Approach

Surgical approach to the hernia in the present case was abdominal: What are some advantages or disadvantages of an abdominal as compared to a thoracic approach? The abdominal procedure allows exploration for additional malformations of the gastrointestinal tract. Normally, the abdominal cavity needs to be stretched to accommodate the herniated intestine; this is best done utilizing an abdominal incision or by attachment of a Silastic sheet (nylon sheet covered on both sides with a layer of silicone rubber) to the edge of the abdominal incision if the herniated contents are too large to fit into the abdominal cavity. In neonates an abdominal approach is practical to determine whether the gut will fit into the abdomen; however, a surgical approach through the thorax immediately relieves pressure on the lungs, and thoracic malformations as well as a hernial sac, if present, are recognized. Closure of a thoracic incision is easy and secure, with fewer wound complications than with an abdominal approach. If necessary, a thoracic incision can be extended to the abdomen.

Associated Malformations

What malformations are most commonly associated with a postero-lateral diaphragmatic hernia? Abnormal fixation or rotation of the gut and hypoplasia of the lung on the involved side are closely correlated with a diaphragmatic hernia. Malformations less frequently seen are gross defects of the neural axis, skull, and vertebral column.

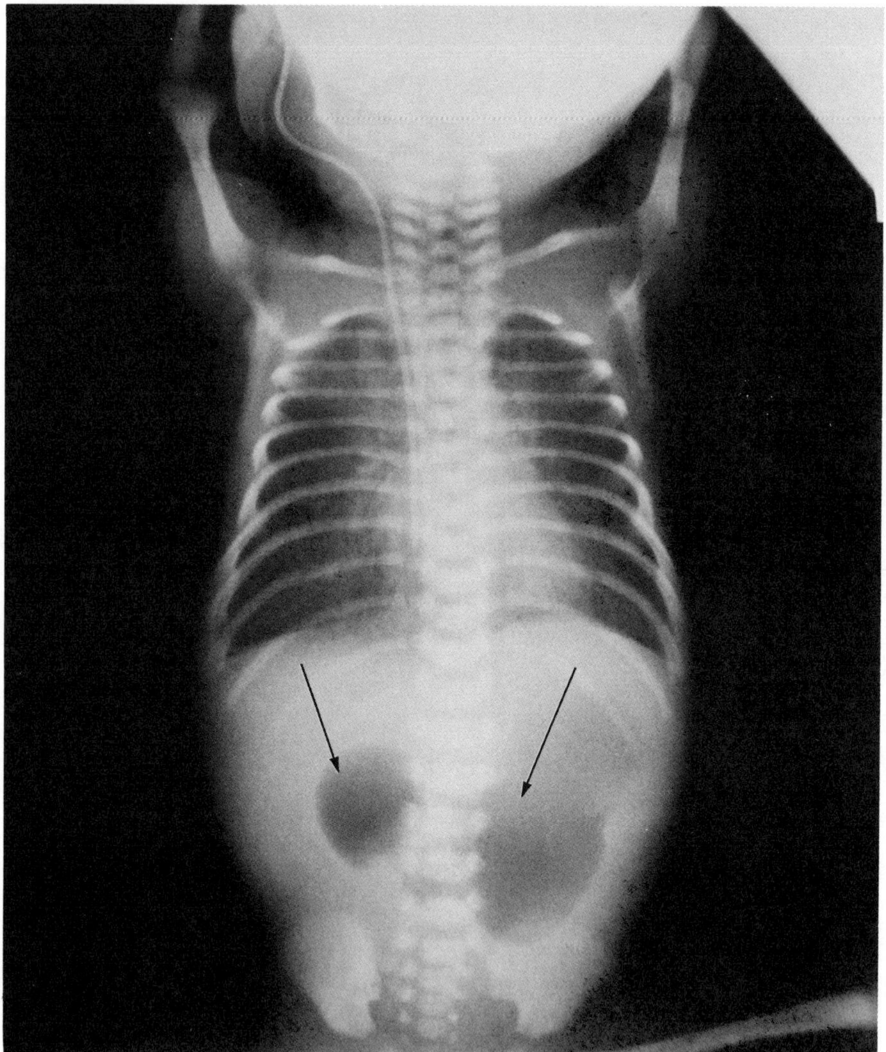

X-ray picture of an annular pancreas showing gas in the duodenal bulb and the third part of the duodenum — "double bubble," *arrows*.

19

Annular Pancreas

A full-term male infant with suggested mongoloid characteristics was born of a pregnancy complicated by maternal polyhydramnios. Initial feedings were accompanied by forceful vomiting of bile-stained material. All subsequent attempts at oral feedings also were accompanied by vomiting and by the third day the infant was mildly jaundiced. X-ray examination without contrast media (plain film) revealed gas in the duodenal bulb and the third part of the duodenum ("double bubble").

DIAGNOSIS

A tentative diagnosis of congenital duodenal stenosis or annular pancreas was suggested, rather than a peptic ulcer, because of the age of the patient.

TREATMENT

Surgical intervention revealed pancreatic material encircling the second part of the duodenum distal to the ampulla of Vater. A bypass procedure, duodenojejunostomy, was employed, leaving the pancreatic tissue intact. This method has the advantage of preventing stasis in the duodenum above the obstruction. The remainder of the intestine was examined for additional stenosis. No associated major malformations were disclosed; the prognosis was excellent, and by the sixteenth day following surgery the infant had adapted well to oral feeding.

DISCUSSION

Annular pancreas is a developmental malformation occurring in the second part of the duodenum and is one of the few congenital defects of the gastrointestinal tract that can produce symptoms in both infants and adults. Extrinsic obstruction of the duodenum is often attributed to an annular pancreas, but strictly speaking this is not entirely correct, as the lumen of the duodenum within the collar of the annual pancreas is not only compressed by the pancreatic tissue, but exhibits an all-

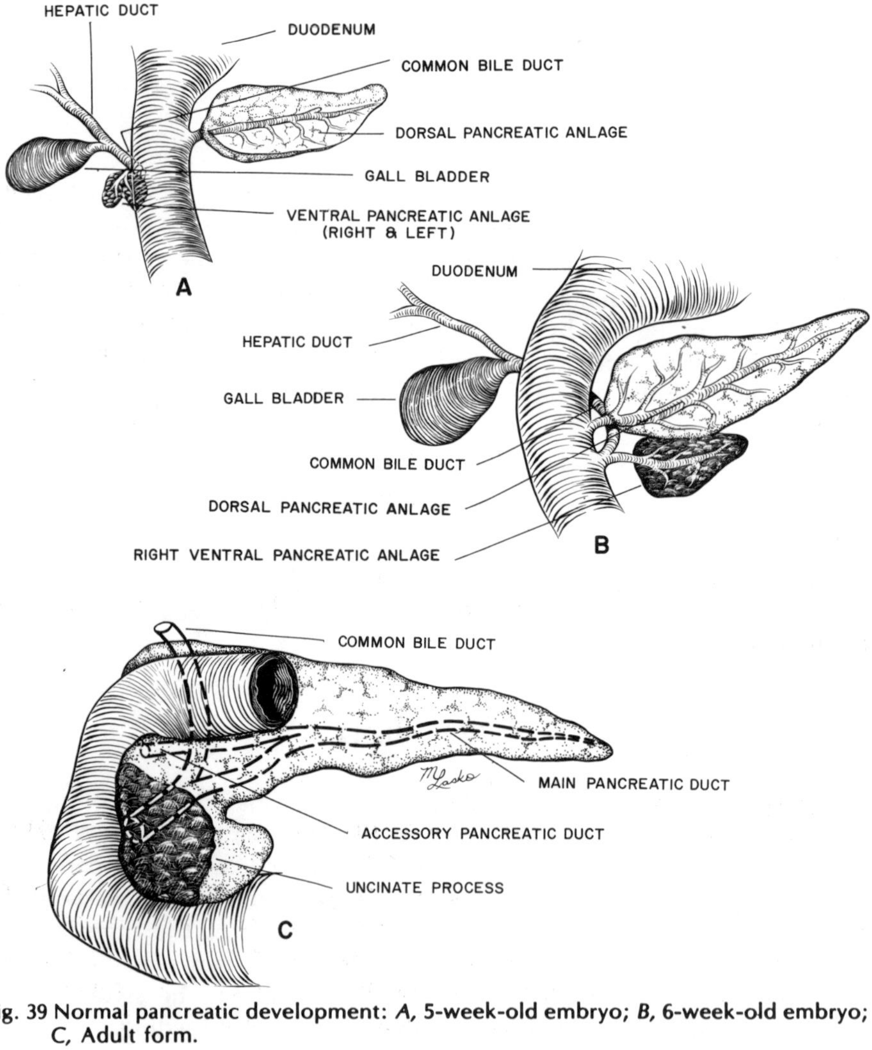

Fig. 39 Normal pancreatic development: ***A,*** **5-week-old embryo;** ***B,*** **6-week-old embryo;** ***C,*** **Adult form.**

around narrowing. This defect in the adult was initially described by von Tiedemann (1818)[31]; von Ecker (1862)[10] introduced the term annular pancreas, while Vidal (1905)[34] described successful surgery in the infant.

Embryology

What is the developmental sequence of the normal pancreas? The pancreas develops during the second month of gestation from dorsal and ventral primordia (anlagen) (Fig. 39A). The dorsal primordium, the first to arise, lies at the same level as the hepatic diverticulum and gives rise to the entire pancreas (tail, body, and superior portion of the head) except the lower part of the right quadrant (uncinate process) of the head (Fig. 39C). The ventral primordium (Fig. 39B) forms two segments (right and left) which grow caudally in the angle between the duodenum and hepatic diverticulum. The left segment becomes atrophic and disappears. As the intestine rotates, the residual right segment of the ventral primordium moves dorsally and to the right and fuses with the dorsal primordium. The ventral primordium forms the lower right quadrant (unicinate process) of the head of the pancreas (Fig. 39C).

Theories

How do you explain embryologically the formation of an annular pancreas? At the present time, there are several theories:

1. *Tieken*[32] proposed that the dorsal and ventral primordia do not fuse, but develop independently on either side of the intestine. As growth proceeds, the bowel is surrounded by pancreatic tissue (Fig. 40A).
2. *Lecco*[18] suggested that if the cranial portion of the right segment of the ventral primordium becomes fixed to the duodenum at the time of intestinal rotation, a band-shaped organ is formed around the duodenum (Fig. 40B).
3. *Baldwin*[2] presumed that the ring of pancreatic tissue was either a persistence of the left segment of the ventral primordium or an excessive growth from the right segment of this primordium. If the latter condition occurs, excessive growth takes place ventrally and to the left of the duodenum at the same time that pancreatic tissue is growing or being carried dorsally to ultimately fuse with the cranial portion of the head of the pancreas (Fig. 40C).

4. *Erimoglu*[11] maintained that the primitive intestine retains the potential ability to form pancreatic tissue after fusion of the dorsal and ventral primordia, and that normally this potential is not restricted to these primordia. Accessory pancreatic tissue may appear and fuse to produce a ring-form pancreas.

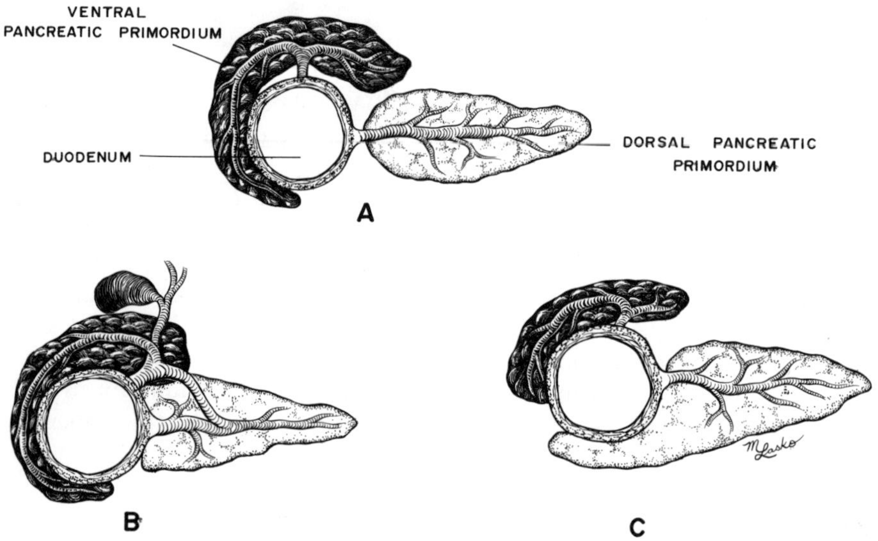

Fig. 40 Theories of annular pancreas development: *A*, Teiken's; *B*, Lecco's; *C*, Baldwin's.

Intestinal Obstruction

What are some of the cardinal manifestations of congenital intestinal obstruction?[5] Abdominal distention, bilious vomiting, abnormality of the meconium evacuation, and maternal polyhydramnios should be used by the physician as indicators of intestinal obstruction in the newborn infant.

Clinical Types

The clinical picture of annular pancreas may be divided into three types:

1. Duodenal obstruction is nearly complete and surgical intervention within a few days after birth is necessary.
2. Duodenal obstruction does not occur until relatively late in life.
3. Duodenal obstruction is minimal, or does not occur, and the annular pancreas is discovered incidentally at necropsy.

Associated Malformations

What malformations are closely associated with annular pancreas? Mongolism (Down's syndrome) is the most frequently associated defect. In addition, atresia or stenosis of the intestine, esophageal atresia, malrotation of the gut, Meckel's diverticulum, and heart defects are sometimes found in combination with an annular pancreas.

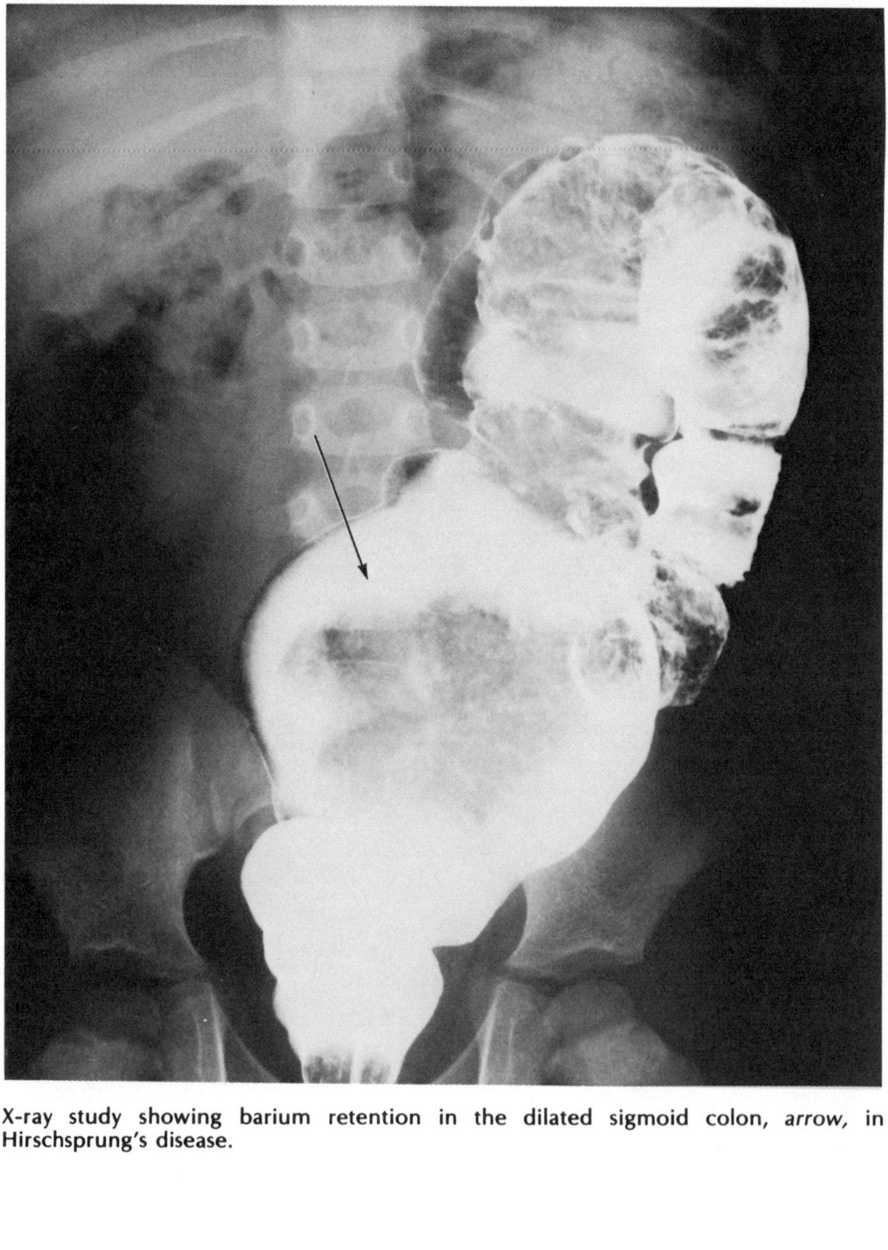

X-ray study showing barium retention in the dilated sigmoid colon, *arrow,* in Hirschsprung's disease.

20

Hirschsprung's Disease

A full-term male infant (3,400 gm.) was delivered following an uncomplicated pregnancy of a gravida II, para II mother. Initial physical examination indicated a normal infant, but there was a delay in the passage of meconium and by the third day there was progressive abdominal distention accompanied by bile-stained vomitus. The infant was transferred to a special-care ward. A previous male sibling had died at 2 months of age of what was described as intestinal obstruction.

EXAMINATION

Rectal examination induced passage of meconiun, followed by visible deflation of the distended abdomen; this proved temporary, however, and the distention recurred within 24 hours. Barium enema revealed a dilated sigmoid colon and a collapsed distal rectal segment. A film taken 24 hours later disclosed barium retention in the dilated colonic segment. Rectal biopsy confirmed the absence of ganglia, which were replaced by clusters of enlarged unmedullated nerve fibers in the submucosal (Meissner's) and intermuscular myenteric (Auerbach's) plexuses of the rectum.

DIAGNOSIS

Short-segment congenital aganglionic megacolon (Hirschsprung's disease, hypoganglionosis, immature ganglia, segmental dilatation, congenital intestinal aganglionosis).

Conservative treatment utilizing daily rectal wash-outs of normal saline yielded unsatisfactory results, and a colostomy through the left iliac fossa was performed on the normal segment of dilated sigmoid colon just proximal to the collapsed aganglionic rectal segment. After

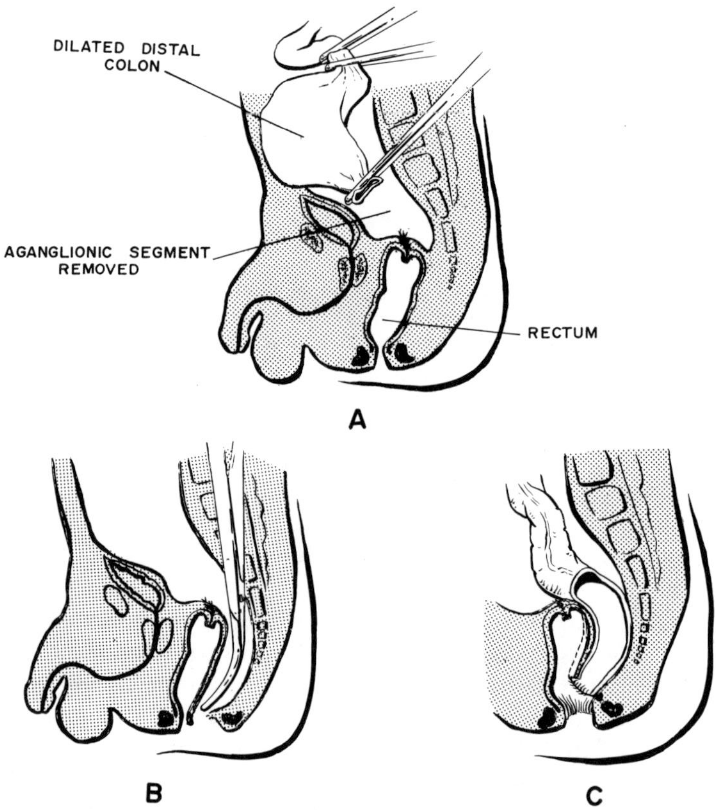

Fig. 41 Modified Duhamel operation: *A*, The dilated distal colon is resected, aganglionic segment removed, and the rectum closed at the level of the pelvic floor; *B*, Posterior rectal wall is incised above the pectinate line; *C*, Colon is "pulled through" incision and re-established with lower rectum. (From Sieber, W. K., and Kieswetter, W. B.: Arch. Surg. 87:111, 1963. Copyright 1963, American Medical Association, with permission.)

two years the colostomy was closed, and a rectosigmoidectomy using a modified Duhamel technique (Fig. 41) was performed by joining the proximal ganglionated bowel to the rectum using a side-to-side anastomosis between fibers of the internal anal sphincter. At 5 years of age, the patient showed normal growth and development.

DISCUSSION

In 1888, Hirschsprung[16] described the clinical and autopsy findings in two infants who had died of constipation associated with dilatation and hypertrophy of the colon. He was the first to recognize that a congenital malformation might be the cause of this disease, which occurs in 1 of 25,000 live births.[25] Hirschsprung's disease may occur as a short- or long-segment type and 90 per cent of the short-segment involvement occurs in infants.[24, 29] By 1901 Tittel[33] had reported cellular degeneration of Auerbach's and Meissner's plexuses distal to the dilated segment in the apparently normal colon. Swenson and Bill[30] in 1948 performed the first curative operation for megacolon utilizing a "rectosigmoidectomy."

Embryology

How do you explain embryologically the presence of a congenital megacolon? Those structures embryologically derived from the hindgut (distal transverse colon, descending colon, sigmoid colon, rectum, and part of the anal canal) are innervated predominantly by parasympathetic fibers via the pelvic splanchnic nerves, and it was suggested that failure in development of these pelvic splanchnic nerves might produce a congenital megacolon. Bodian[4] conjectured that a genetically determined congenital megacolon resulted from a failure of the cells of the dorsal crest of the neuroectoderm to migrate in a normal craniocaudal direction; an early disturbance would cause a long aganglionic segment, a later disturbance would produce a short aganglionic segment. Wildegans[36] showed that the ganglion cells within the wall of the intestine differentiate between the seventh and eighth week of gestation in a craniocaudal direction independent of any parasympathetic nerve supply. Studies by Okamoto and Ueda[22] indicated that the ganglion cells enter the cranial part of the intestine during the sixth or seventh week of gestation, probably via the vagus nerve, and migrate caudally between the muscle layers of the intestinal tube, with the myenteric plexus maturing several weeks earlier than the submucosal plexus. From these theories, it appears that in Hirschsprung's disease, due to arrested development, the course of caudal migration of the ganglion cells is interrupted.

In the neonate, what intestinal malformations or syndromes may mimic aganglionosis and must be considered in a differential diagnosis of congenital megacolon?[24, 28] Meconium plug syndrome, inspissated milk syndrome, immaturity of bowel functions, necrotizing enterocolitis, meconium ileus, intestinal malrotation, and lower ileal atresia must be considered in a differential diagnosis.

Appearance of Gut

Describe the histological appearance of the segments of gut involved in the present case of congenital megacolon:

1. Dilated proximal segment of megacolon (distal sigmoid colon). Normal ganglion cells are present.
2. Junctional zone (cone) of gut between sigmoid colon and rectum. The few ganglion cells present are abnormal in appearance.
3. Collapsed distal segment of megacolon (proximal rectum). Ganglion cells are absent and replaced by an increased number of clusters (whorls) of enlarged unmedullated nerve fibers.

Methods of Treatment

What are the major methods of treating congenital megacolon? Surgery has been the most successful treatment since 1948 when Swenson and Bill introduced "rectosigmoidectomy" (pull-through), a technique most suitably used in children past infancy. A modification of the Swenson and Bill technique by Duhamel[9] is more frequently employed on infants, as in the present case. Recently, Soave[27] introduced an abdominoperineal approach which avoids pelvic dissection. The normal segment of colon is carried through a portion of aganglionic rectum and united to the anal sphincter, forming an anocolic anastomosis.

Associated Malformations

What malformations might be associated with congenital aganglionic megacolon? Megalobladder and megaloureter occasionally accompany a congenital megacolon. Additional abnormalities include mongolism, cardiac malformations, malrotation or nonrotation of the midgut, Meckel's diverticulum, and congenital short bowel.

REFERENCES

1. Ahlfeld, F.: Der Alkohol bei der Behandlung inoperabeler Bauchbruche. Wschr. Geburtsh. Gyn. 10:24, 1899.
2. Baldwin, W. M.: A specimen of annular pancreas. Anat. Rec. 4:299, 1910.
3. Bochdalek, V.: Einige Betrachtungen ueber die Entstchung des

angeborenen Zwerchfellbruches: Als Beitrag zur pathologischen Anatomie der Hernien. Vsb. prakt. Heilk. 19:89, 1848.

4. Bodian, M., and Carter, C. O.: A family study of Hirschsprung's disease. Ann. Hum. Genet. 26:261, 1963.
5. Clathworthy, W. H., Jr., and Lloyd, J. R.: Intestinal obstruction of congenital origin. Arch. Surg. 75:880, 1957.
6. Cooper, Sir Astley: Anatomy and Surgical Treatment of Abdominal Hernia. Lea and Blanchard, Philadelphia, 1844.
7. Creighton, R. E., Whalen, J. S., and Conn, A. W.: The management of congenital diaphragmatic hernia. Can. Anaesth. Soc. J. 13:124, 1966.
8. Davis, R. M., and Kehm, R. W.: Omphalocele with patent vitellointestinal duct and ileal prolapse. Am. J. Surg. 113:571, 1967.
9. Duhamel, B.: Une nouvelle operation pour le megacôlon congenital. Presse Med. 64:2249, 1956.
10. von Ecker, A.: Bildungsfehler des Pancreas und des Herzens. Medicin. 14:354, 1862.
11. Erimoglu, C.: A case of pancreas anulare. Proc. Kon. Nederl. Akad. Wet. [Biol. Med.] 55:18, 1952.
12. Gennaro, A. R., and Bacon, H. E.: The merit of selectivity in the surgical management of congenital megacolon. Dis. Colon Rectum 11:91, 1968.
13. Grob, M.: Conservative treatment of exomphalos. Arch. Dis. Child. 38:148, 1963.
14. Gross, R. E.: Congenital hernia of the diaphragm. Am. J. Dis. Child. 71:579, 1946.
15. Gross, R. E.: A new method for surgical treatment of large omphaloceles. Surgery 24:277, 1948.
16. Hirschsprung, H.: Stuhlträgheit Neugeborener in Folge von Dilatation and Hypertrophie des Colons. Jahrb. Kinderheilk. 27:1, 1888.
17. Johnson, T.: The workes of that famous Chirurgion Ambrose Parey. Translated out of Latine and compared with the French. Th. Cotes and R. Young, London, 1634.
18. Lecco, T. M.: Zue Morphologie des Pancreas annulare. Sitzungsbericht. Wien. Akad. Wiss. 119:391, 1910.
19. Meckel, J. F., the younger: Handbuch der Pathologischen Anatomie. C. H. Reclam, Leipzig, 1812-1816.
20. Millard, R., Pigott, R., and Zies, P.: Free skin grafting of full-thickness defects of abdominal wall. Plast. Reconstr. Surg. 43:569, 1969.

21. Mustard, W. T., Ravitch, M. M., Snyder, W. H., Jr., Welch, K. J., and Benson, C. D.: Pediatric Surgery. Year Book Medical Publishers, Chicago, 1969.
22. Okamoto, E., and Ueda, T.: Embryogenesis of intramural ganglia of the gut and its relation to Hirschsprung's disease. J. Pediat. Surg. 2:437, 1967.
23. Olshausen, R. Z.: Zur Therapie der Nabelschnurhernien. Arch. Gynaek. 29:443, 1887.
24. Rickham, P. P., and Johnston, J. J.: Neonatal Surgery. Appleton-Century-Crofts, London, 1969.
25. Rubin, A.: Handbook of Congenital Malformations. W. B. Saunders Company, Philadelphia, 1967.
26. Schuster, S. R.: A new method for the staged repair of large omphaloceles. Surg. Gynec. Obstet. 125:837, 1967.
27. Soave, F.: A new surgical technique for treatment of Hirschsprung's disease. Surgery 56:1015, 1964.
28. Seminar on Pseudo-Hirschsprung's Disease and Related Disorders. Combined meeting of the British Association of Paediatric Surgeons and the Surgical Section of the American Academy of Pediatrics. Edinburgh, July 27-30, 1965. Arch. Dis. Child. 41:143, 1966.
29. Soderland, S.: Meckel's diverticulum, a clinical and histologic study. Acta Chir. Scand. Suppl. 248:1, 1959.
30. Swenson, O., and Bill, A. H.: Resection of rectum and rectosigmoid with preservation of the sphincter for benign spastic lesions producing megacolon. Surgery 24:212, 1948.
31. Tiedemann, F. von.: Ueber die Verschiedenheiten des Ausführungsgangs der Bauchspeicheldruse bei den Menschen und den Säugethieren. Deutsch. Arch. Physiol. 4:403, 1818.
32. Tieken, T.: Annular pancreas. Amer. Med. 2:826, 1901.
33. Tittel, K.: Ueber eine angeborene Missbildung des Dickdarmes. Wien. Klin. Wschr. 14:903, 1901.
34. Vidal, E.: Pancréas annulaire et atrophie du duodenum chez un enfant de trois jours. Gastro-entérostomie postérieure. Guérison. Dixhuitième Congrés de Chirurgie, Paris. 18:745, 1905.
35. Wesselhoeft, C. W., and Randolph, J. G.: Treatment of omphalocele based on individual characteristics of the defect. Pediatrics 44:101, 1969.
36. Wildegans, H.: Die Krankheiten und Verletzungen des Dickharms und des Mastdarms. Enke Verlag, Stuttgart. 1969.

PELVIS

21

Undescended Testis

Scrotal examination of a full-term male infant revealed an empty scrotal sac on the right side. Palpation of the groin and inguinal area located the testis in the superficial inguinal pouch.[4] The left testis was in the normal position within the scrotum.

DIAGNOSIS

Right unilateral undescended testis (improperly descended testis, cryptorchidism, maldescent of the testis).

FURTHER COURSE

When the infant was placed in a warm bath, the scrotal sac on the right side filled out, a good indication that in time the testis would permanently descend into the scrotum. Since the left testis was in the normal position, fertility was assured and hormone treatment (chorionic gonadotrophin) to initiate descent of the right testis was withheld. An undescended testis of this type usually completes its descent a few weeks after birth or at puberty. If at puberty the testis is not descended, orchiopexy is considered. If the length of the spermatic cord is insufficient to allow placement of the testis in the scrotum, the testis may be removed. The precise treatment will vary with the case and the physician.

DISCUSSION

The intra-abdominal position of the fetal testes was known to Galen.[14] Cabral (1564)[17] described bilateral absence of the testes, while

John Hunter (1762)[19] gave the earliest documented account of undescended testis, and in a 7-month fetus located the testes in the abdomen and proposed that they are directed in their descent by the gubernaculum.

Embryology

Discuss the origin and the descent of the testes into the scrotum. The gonads are intimately associated with the nephric system from their earliest appearance in the 4-week embryo (Fig. 42). The gonads arise

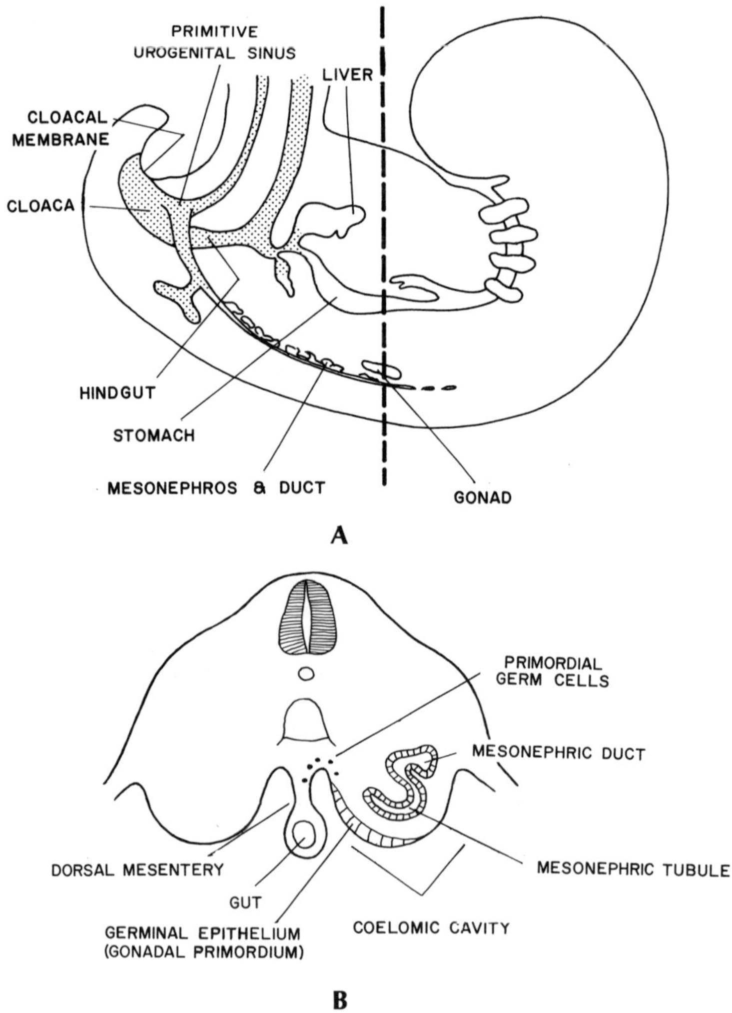

Fig. 42 *A*, Development of the gonad. *B*, Cross-section at level of developing gonad.

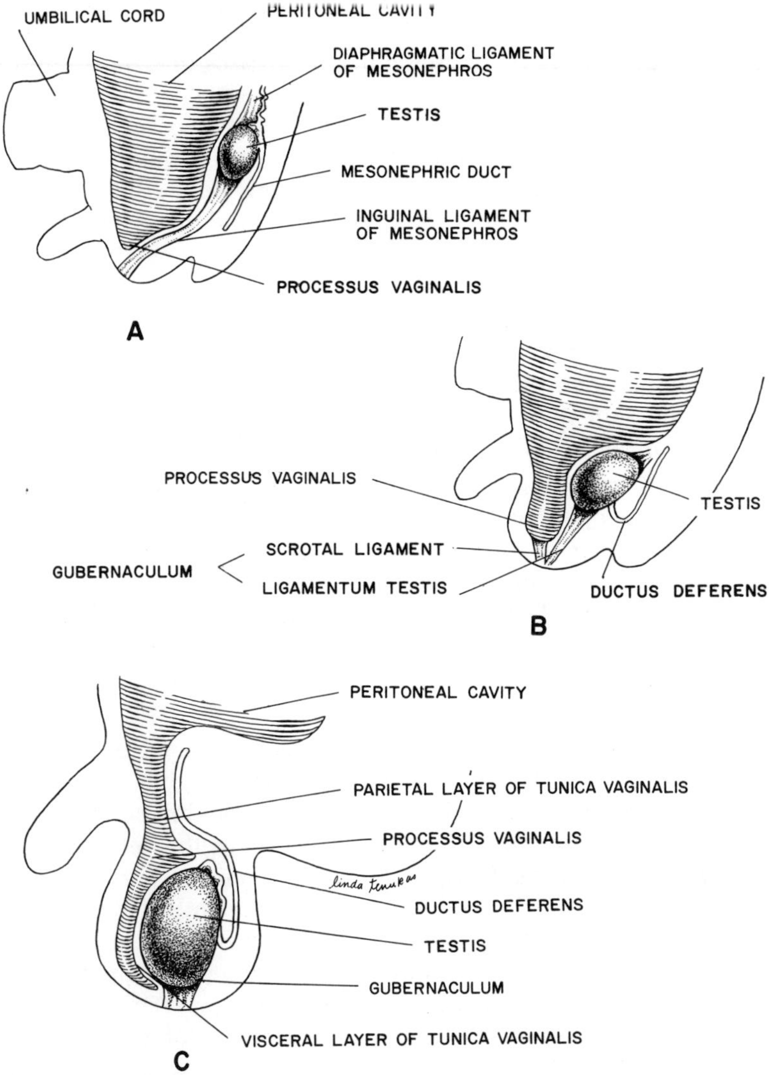

Fig. 43 Descent of the testes: *A,* 12-week-old fetus; *B,* 20-week-old fetus; *C,* 32-week-old fetus.

as thickenings (gonadal ridges) on the ventral and medial borders of the mesonephros (Fig 42 B). The mesonephros grows rapidly and bulges into the coelom, pushing ahead of it a covering of peritoneum which extends over the developing gonad. At the cranial and caudal ends of the mesonephros, the peritoneum elongates into two folds. The superior fold is the diaphragmatic ligament of the mesonephros, while the inferior fold becomes fibrous and is known as the inguinal ligament of the mesonephros. As the mesonephros decreases in size, the testis enlarges to incorporate more of the peritoneal covering of the mesonephros. Simultaneously, the testis becomes closely situated to the inguinal ligament, which expands and is gradually taken over by the growing gonad. The inguinal ligament then becomes the ligamentum testis. As the abdominal cavity increases in size, each testis and its associated structures remain attached to the anterior wall by the ligamentum testis.

In the third month (Fig. 43A) the coelomic cavity begins to form an evagination on each side of the midline into the scrotal swellings. The extension of the coelomic cavity into the scrotal swellings is the processus vaginalis, and the connective tissue surrounding this processus extends to the skin of the scrotum to form the scrotal ligament. The scrotal ligament and the ligamentum testis constitute the gubernaculum.

In the fifth month (Fig. 43B) the processus vaginalis reaches the scrotum and the opening from the coelom into the scrotum becomes smaller. The margins of this opening are reinforced by connective tissue to form the internal inguinal ring. The definitive scrotal pouch is lined with the peritoneum of the processus vaginalis and covered externally by skin. An extension of the muscular and fascial layers of the anterior abdominal wall is found in the scrotum between the peritoneal lining and the skin.

With the formation of the scrotal pouches, the testes and the retained mesonephric structures, the latter of which give rise to part of the testicular duct system, descend to a more caudal position, and by the end of the fifth month the testes lie close to the internal inguinal ring. During the seventh month the testes begin to pass through the inguinal canal, and by the end of the eighth month (Fig. 43C) they normally reach the scrotal pouches. The testes remain at the apex of the gubernaculum; moreover, failure of the gubernaculum to elongate in proportion to the growth of the pelvis is partly responsible for the change from the intra-abdominal to scrotal position of the testes.

The testis moves caudally beneath the peritoneal lining of the processus vaginalis in its descent. Therefore, it does not enter the lumen of the scrotal pouch directly, but slips under the peritoneal lining and protrudes into the lumen, reflecting the peritoneal layer over itself. This peritoneal layer becomes the visceral layer of the tunica vaginalis proprius. The remainder of the peritoneal sac that lines the scrotal cavity becomes the parietal layer of the tunica vaginalis proprius (Fig. 43C).

What embryological structure may be responsible for the various positions of an undescended testis? Lockwood[26] reports that the distal part of the gubernaculum may divide into several tails: (1) scrotal; (2) perineal; (3) pubopenile; and (4) femoral. After passing the external inguinal ring, the pathway the testis takes may be determined by the tail of the gubernaculum, which keeps open and dilates this pathway and the scrotum. Moreover, the gubernaculum permits testicular descent by its differential growth, but does not pull the testis into the scrotum.

Classification

How are the various types of maldescended testes classified? Im-

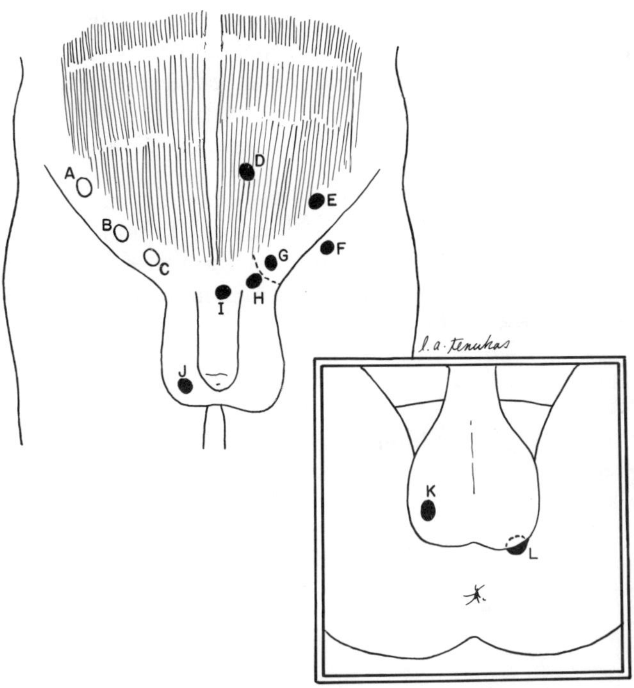

Fig. 44 Maldescent of the testes. Malposition along normal descent pathway: *A*, Renal; *B*, Deep inguinal; *C*, Superficial inguinal; *D*, Superficial abdominal; *E*, Superficial inguinal pouch; *F*, Femoral; *G*, Superficial inguinal; *H*, Pubic; *I*, Penile; *J* and *K*, Scrotal; *L*, Perineal. (From Hamilton, W. J., Boyd, J. D., and Mossman, H. W.: Human Embryology, ed. 3. W. Heffer & Sons, Cambridge, England, 1964, with permission.)

properly descended testes (Fig. 44), which occurs in 1/500[40] males, is categorized as follows:

1. *Rectractile testes* (migratory)—the powerful contraction of the cremasteric muscle causes the testes to reside principally in the superficial inguinal pouch. The testes can be manipulated to return to the scrotum.
2. *Ectopic testes*—the testes complete normal migration through the inguinal canal, but do not enter the scrotum. The testes may be located in the: (a) perineal; (b) pubopenile; (c) femoral; or (d) pubic position; or in the (e) superficial inguinal pouch.
3. *Undescended testes* (retained)— the testes do not descend completely through the inguinal canal, but may take one of three positions: (a) they are retracted in the young child, but reside in a normal scrotal position at puberty; (b) descent is obstructed, but the histology of the testes is normal; and (c) the histology of the testes is abnormal and they may be located in the abdomen, inguinal canal, or suprascrotal area.

Complications

What are the common complications of undescended testes? Sterility may result from maldescent of both testes. Inguinal hernias, trauma, emotional disturbances, and malignancy of the testes are other complications; testicular malignancy probably occurs more often in undescended than in descended testes.[24]

Associated Malformations

There is a high incidence of urinary tract malformations in cryptorchidism.

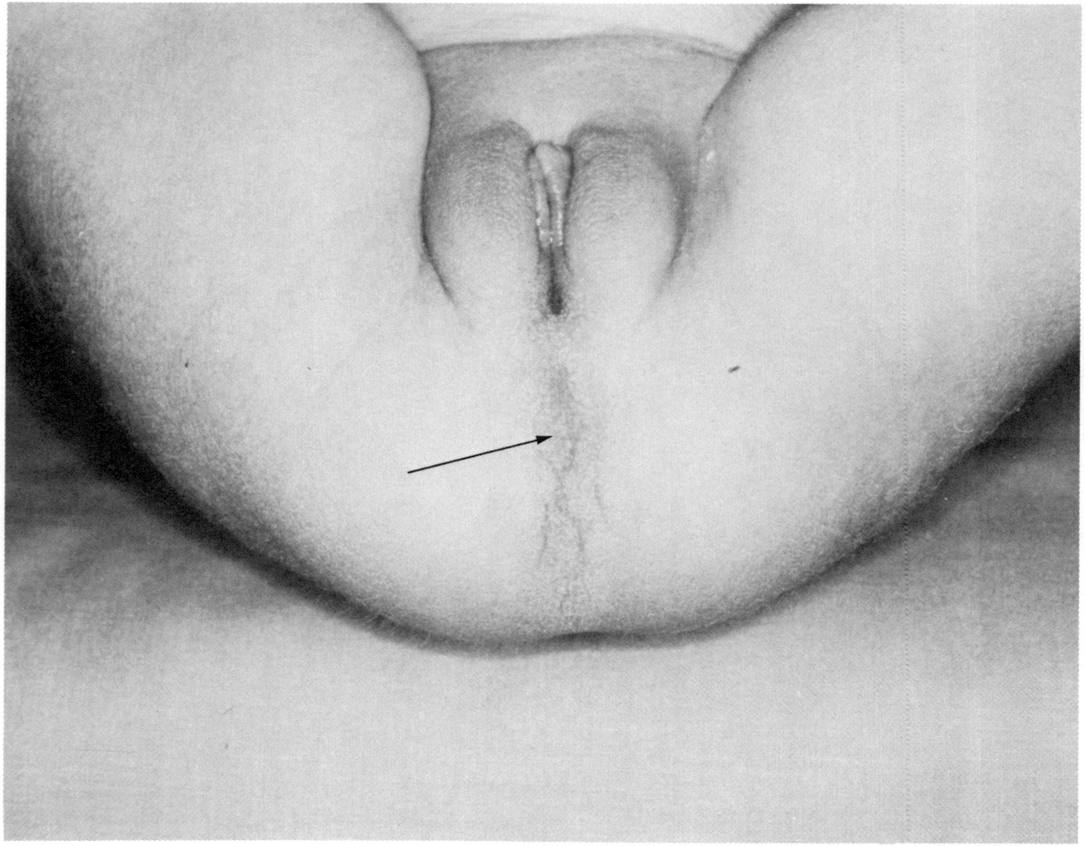

Imperforate anus, *arrow*.

22

Imperforate Anus

Initial inspection of the perineal area of a full-term newborn male infant revealed no anal opening. Pigmented skin surrounded an anal dimple. Upon "thumping" the buttock, the area surrounding the dimple contracted, indicating an intact and functional external anal sphincter.

EXAMINATION

Squamous cells and small amounts of meconium were identified in a urinalysis, signifying a fistular connection between the cul-de-sac of the gut and the urinary system. Further examination was postponed for 24 hours, allowing gas to enter the lowest part of the bowel prior to x-ray evaluation. The infant was placed in an inverted position for three minutes, after which the perineal area was painted with barium paste, the knees flexed, and two lateral pictures made. The x-rays showed the rectum ending in a blind pouch above the pubococcygeal line of Stephens (line from the center of the pubis to the sacrococcygeal junction). Dye injection via the urethra confirmed a fistular connection between the blind rectal pouch and the prostatic urethra.

DIAGNOSIS

Imperforate anus (type III, high lesion) with rectourethral fistula.

TREATMENT

No additional malformations were noted and surgery utilizing an abdominoperineal approach was immediate. Mobilization of the bowel and division of the fistular connection was made through the abdomen. An incision in the anal dimple allowed the mobilized rectum to be pulled through this newly formed opening. Care was taken to pass the rectum through the puborectalis muscle, which acts as a sling support, and to preserve sensory and motor nerves to the levator ani, rectum, and genitourinary systems. Four days after surgery a normal bowel movement occurred, and by the tenth day regular dilatations of the newly formed anus were begun and continued at home by the mother. When the infant was 1 year old, fecal and urine continence appeared normal and bowel movements were regular. The complete success of this repair cannot be evaluated until toilet training is begun at 2 to 2½ years.

DISCUSSION

Imperforate anus, which occurs in 1/5,000 births,[40] was recognized by early Greek, Roman, and Arabic physicians. In the seventh century A.D. a Byzantine physician, Paulus Aegineta,[30] was credited with the first mention of surgical intervention, while Littré (1710)[25] suggested the use of colostomy as a palliative procedure. In 1835, Amussat[1] made one of the most significant contributions to this field by describing a method of perineal dissection to locate the rectal pouch. A combined abdominoperineal approach for repair of an imperforate anus was initiated in 1948 by Rhodes, Pipes, and Randal.[39]

Embryology

Explain the formation of the rectum. In the present case study, how do you account for the imperforate anus?

Embryonic changes from the third to ninth week of intrauterine life transform a common chamber (cloaca) into a urogenital portion and a digestive portion.

By the end of the third week (Fig. 42), the primitive urinary bladder, which is formed by an expansion of the allantois and which receives the paired mesonephric ducts, and the hindgut empty into the primitive cloaca. The cloaca itself is separated from the amniotic cavity by a cloacal membrane composed of entoderm and ectoderm.

At approximately the sixth week (Fig. 47B), the urorectal septum, a mesodermal mass covered with peritoneum, begins to descend inferiorly and ventrally, resulting in the formation of a ventral urogenital sinus and a dorsal primitive rectum; moreover, the communication between the bladder and the hindgut is narrowed to a small cloacal duct.

By the end of the seventh week (Fig. 50D), separation of the urogenital passage from the rectal passage is completed by fusion of the urorectal septum with the cloacal membrane; the latter is then temporarily divided into a urogenital and an anal membrane. Rapid proliferation of mesenchyme beneath the ectoderm surrounding the cloacal membrane forms the genital swellings. Thus, the cloacal membrane comes to lie at the bottom of a depression (external cloaca) surrounded by the genital swellings. Investigators suggest[15, 42] that the urorectal septum divides the cloaca into a urogenital portion and a rectal portion only up to the point of entrance of the paramesonephric (müllerian) ducts into the posterior wall of the urogenital sinus. The remainder of the urorectal septum (from the entrance of the paramesonephric ducts to the cloacal membrane) is formed by lateral invaginations (folds of Rathke) from the wall of the definitive urogenital sinus.

Prior to loss of the anal membrane (Fig. 45A), two small eminences that are connected by a medial band appear externally and

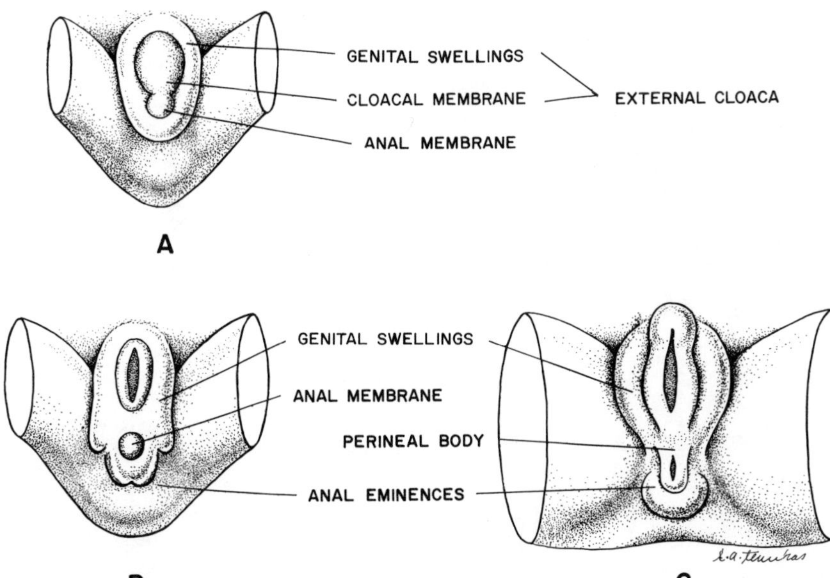

Fig. 45 Division of external cloaca: *A,* 7-week-old embryo; *B,* 8-week-old embryo; *C,* 9-week-old fetus.

dorsally on either side of the future anus and ventral to the tail of the embryo (Fig. 45B). These eminences elongate in a ventral and then a medial direction to converge at the midline. The genital swellings fuse with these anal eminences at approximately the same time as the urorectal septum fuses with the cloacal membrane. The anal eminences, genital swellings, and urorectal septum fuse to form the perineal body (Fig. 45C). Once the perineal body is formed, the external cloaca is divided into a urogenital and a rectal portion. Continued growth of the anal eminences produces a depression externally; this is the proctodeum and is surrounded by the anal mound. The anal mound is composed of external sphincter fibers.

In the present case study, the imperforate anus is the result of an incomplete urorectal septum. This incomplete septum yields a fistular connection between the rectal cul-de-sac and a portion of the definitive urogenital sinus (prostatic urethra of the male).

Classification

Congenital malformations of the rectum and anus are frequently incorrectly listed under the term "imperforate anus." This group of di-

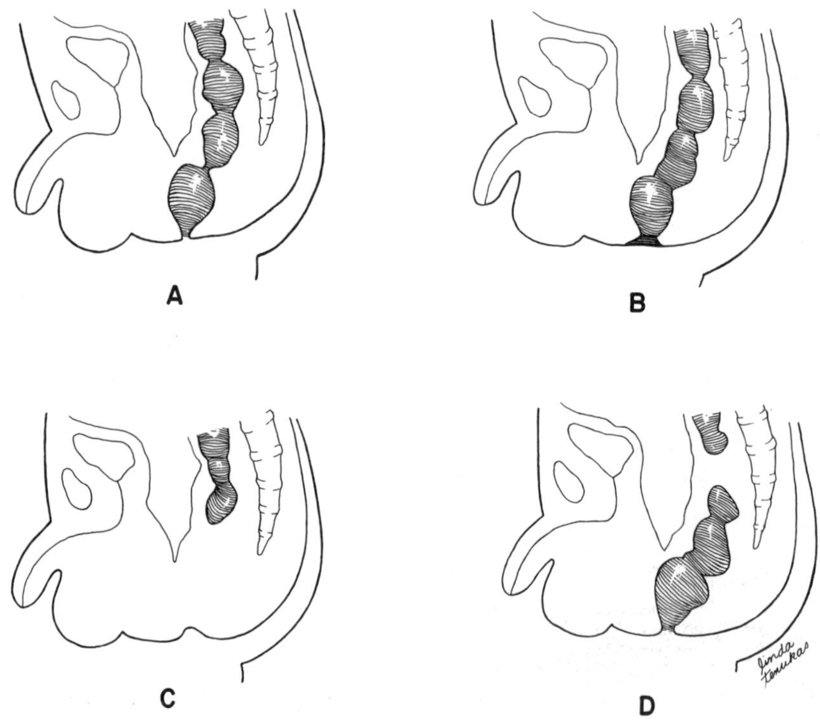

Fig. 46 Classification of anal and rectal malformations: *A,* Stenosis of the anus; *B,* Imperforate anus with separation of rectum from anus; *C,* Membranous obstruction of the anus; *D,* Separation of rectum from anal canal and anus.

verse anomalies includes the "commonest" malformations that are "incompatible" with life. How are these malformations classified? The classification introduced by Ladd and Gross[23] in 1934 is most widely used (Fig. 46), although a number of anomalies do not conveniently fit into this taxonomy. In the present case study, the Ladd and Gross classification with modifications is utilized:

Types of Imperforate Anus

Type I Congenital anal stenosis
Type II Persistent membrane covering the anus
Type III Imperforate anus and rectum in which the rectal pouch terminates blindly some distance above the anus.

A. *High lesions*—The colon terminates above the levator ani. Surgical approach is abdominoperineal.
   1. Anorectal agenesis
   2. Persistent cloaca
   3. Fistulas

B. *Low lesions*—The colon terminates below the levator ani.
   Surgical approach is perineal.
   1. Thick anal covering
   2. Anal stenosis or agenesis
   3. Persistent anal membrane

Type IV Anus and proctodeum are normal, but rectal pouch terminates blindly in the sacrum.

Complications

What postoperative complications may occur after repair of a high lesion as in the present case study? Possible postoperative complications are: mucosal prolapse at junction of rectum and new anal opening; stenosis at surgical junction; separation of the anastomosis; urethral stricture; urethral diverticulum; recurrent rectourinary fistula; urinary or rectal retention or incontinence; and inadequate blood supply to surgical junction.

Associated Malformations

What malformations are most commonly associated with an imperforate anus? The highest percentage of malformations accompanying an imperforate anus are of the urinary and genital systems and remaining portions of the digestive system.

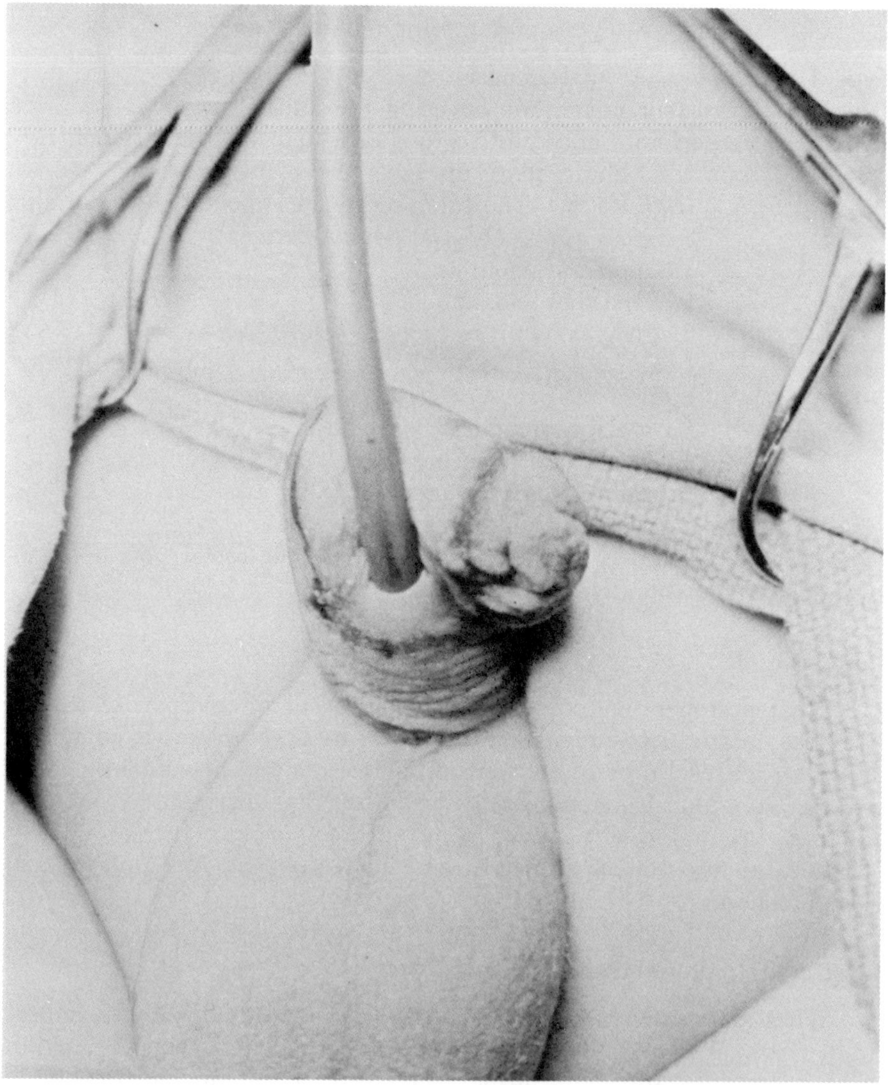

Catheter in the urethra of a penile-type hypospadic.

23

Hypospadias

An 18-year-old primigravid mother delivered a full-term male infant. Routine examination of the infant revealed that the urethral meatal opening terminated ventral and proximal to its normal location at the apex of the glans penis. Redundant dorsal preputial tissue gave the penis a hooded appearance, while the prepuce itself was deficient on the lateral sides and absent on the ventral side of the penis. The prostatic urethra was not involved; thus, normal function of the sphincters and urinary control existed. The right testis was palpated in the superficial inguinal pouch and could be manipulated into the scrotal sac.

DIAGNOSIS

Subcoronal hypospadias with chordee and meatal stenosis and right unilateral undescended testis.

TREATMENT

In repair of hypospadias the prime objectives are (1) to correct the ventral deformity (chordee); (2) to construct a penile urethra at or near the apex of the glans—the construction must be free of hair or scar and capable of normal sexual function; and (3) to utilize or remove the hooded prepuce. In the present case, repair was completed in two stages. When the boy was 2 years of age, the chordee and meatal stenosis were corrected. Due to the straightening of the penis, the urethral opening was now further from the glans than originally. After a lapse of one year to allow the penile skin to regain its normal elasticity and looseness, urethroplasty (stage two) was performed using the Browne technique,[5] in which the urethral tube is lengthened by tubular formation of a buried strip of the prepuce skin.

Several days after surgery, a fistula occurred between the old and new urethral tubes and was promptly repaired. At 5 years of age the boy voided from the tip of the glans with a good urinary stream, and repair was completed prior to entrance into school.

The undescended testis was not treated at this time.

DISCUSSION

Dieffenbach[9] in 1837 was the first to attempt correction of hypospadias, a condition that occurs in 1/300[2] live male births. In 1874, Anger and Mettauer[2] accomplished the first satisfactory correction of both chordee and an abnormally located meatal opening. The successful techniques in hypospadias repair are the result of the early works of Duplay,[10] Ombredanne,[34] and Nové-Josserand.[33] More than 123 surgical techniques and modifications[2, 30] have been utilized for the repair of chordee and hypospadias.

Embryology

Discuss the development of the male urethra. In the four-week-old embryo, the primitive urogenital sinus and the hindgut empty into the cloaca; the latter is separated from the amniotic cavity by the cloacal membrane (Fig. 42). By the end of the seventh week (Fig. 47B), the cloacal membrane fuses with the growing urorectal septum and the membrane then is divided into a urogenital portion and an anal portion. The urogenital portion is separated into: (1) the vesicourethral canal, from which arise the urinary bladder and the upper portion of the prostatic urethra, and (2) the definitive urogenital sinus, which is further divided into pelvic and phallic portions. In the male, the pelvic portion of the sinus becomes the lower part of the prostatic urethra and the membranous urethra. The phallic portion forms the external genitalia with the exception of the urethral portion of the glans penis.

By the end of the seventh week (Figs. 47A, C; 48), a urethral plate develops from entodermal cloacal epithelium related to the base of the genital tubercle. This plate is located immediately above and internal to the cloacal membrane. The urethral plate grows forward into the mesenchyme of the genital tubercle such that the lower margin of this entodermal plate remains against the ectodermal surface epithelium covering the genital tubercle. While the urethral plate is growing for-

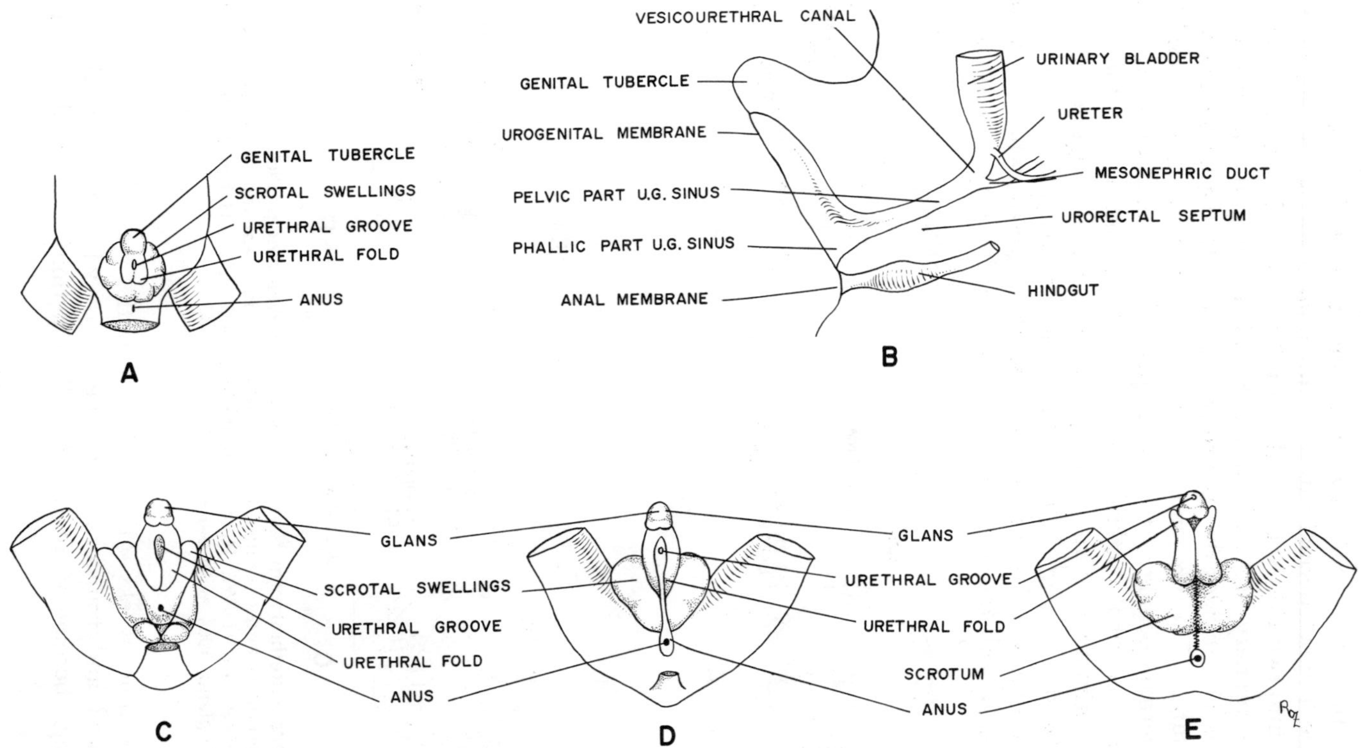

Fig. 47 Male urethral formation: *A,* 5-week-old embryo — ventral view; *B,* 6- to 7-week-old embryo — sagittal section; *C,* 7-week-old embryo — ventral view; *D,* 9-week-old embryo — ventral view; *E,* 12-week-old embryo — ventral view.

ward to reach the tip of the genital tubercle, the mesenchyme on either side of the plate proliferates. Thus, on the ventral aspect of the genital tubercle, the ectodermal surface epithelium covering the mesenchyme is raised into two ridges, called the primitive urethral folds, between which lies the primary urethral groove. Simultaneously, the urorectal septum contacts and divides the cloacal membrane into the urogenital and anal membranes. The urogenital membrane then disintegrates, resulting in the external opening of the phallic part of the urogenital sinus. Presently, the entodermal urethral plate, which is in contact with the ectoderm of the primary urethral groove, thickens and disintegrates along with the related ectoderm. This results in the formation of a deep secondary urethral groove, called the definitive urethral groove, which is continuous proximally with the open phallic portion of the urogenital sinus.

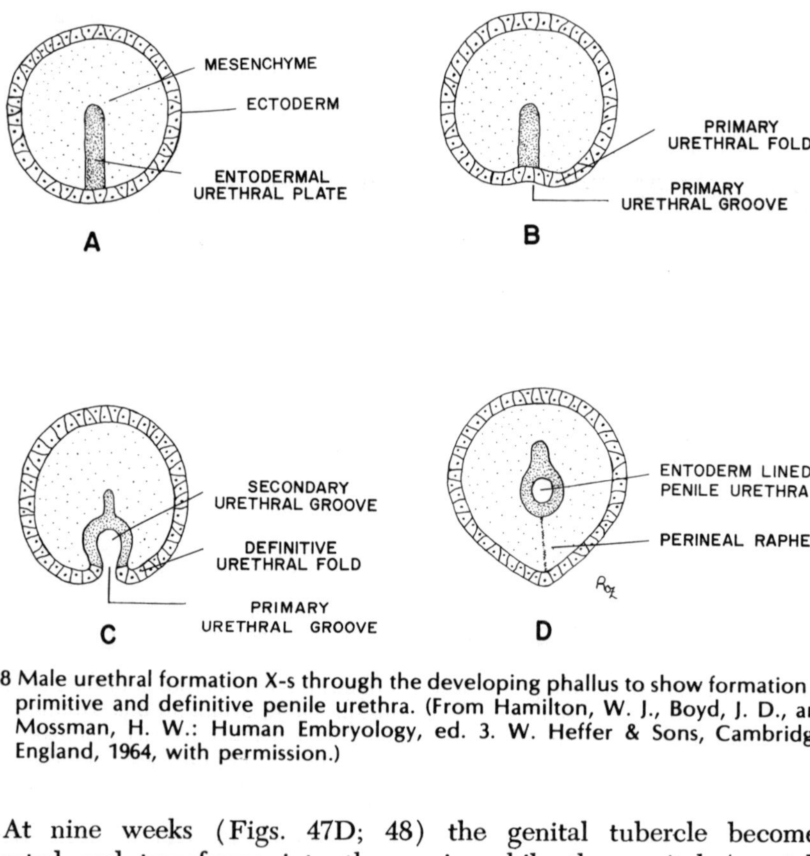

Fig. 48 Male urethral formation X-s through the developing phallus to show formation of primitive and definitive penile urethra. (From Hamilton, W. J., Boyd, J. D., and Mossman, H. W.: Human Embryology, ed. 3. W. Heffer & Sons, Cambridge, England, 1964, with permission.)

At nine weeks (Figs. 47D; 48) the genital tubercle becomes elongated and transforms into the penis, while the genital (scrotal) swellings become more definite. The definitive urethral groove continues forward on the ventral aspect of the penis, but not to the tip. The urethral folds unite and fuse such that only entodermal epithelium from the urogenital sinus (secondary urethral groove) comprises the lining of the penile urethra. The urethra of the glans penis is formed in a manner similar to that described for the penile urethra, but in addition, surface ectoderm grows into the substance of the glans to meet

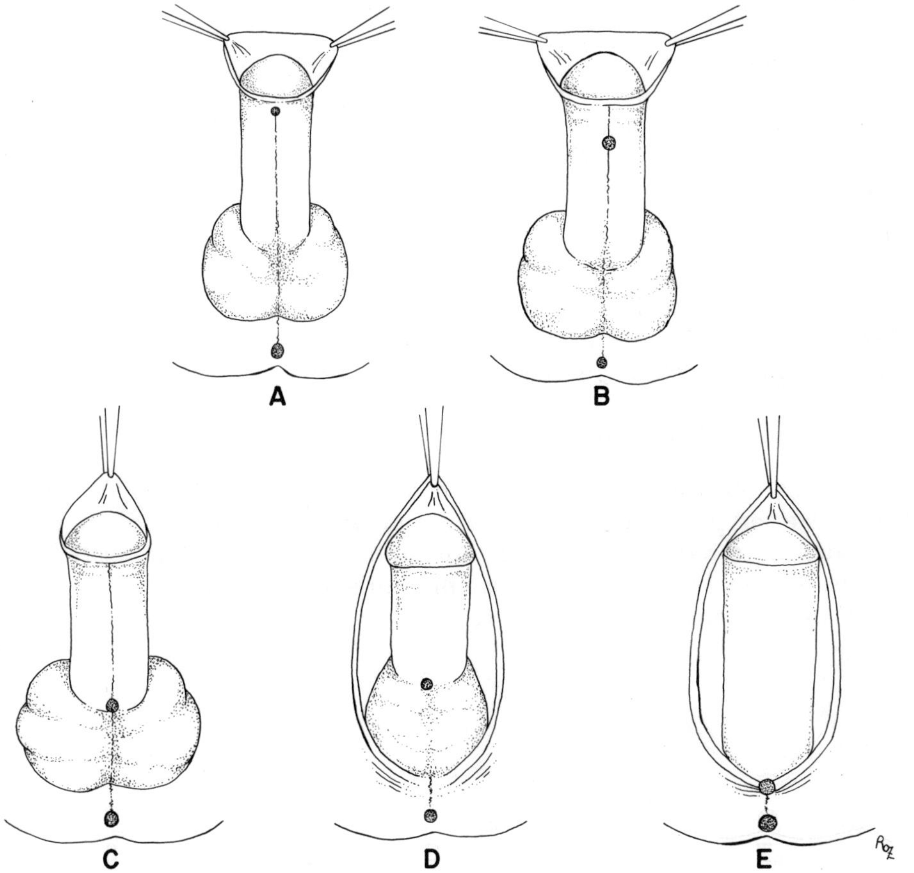

Fig. 49 Forms of hypospadias: *A*, Subcoronal; *B*, Penile; *C*, Penoscrotal; *D*, Scrotal; *E*, Perineal.

the urethral plate, thus giving an ectodermal and entodermal lining to the glans urethra.

In the male, development of the external genitalia is completed by the eleventh week (Fig. 47E).

Explain the development of hypospadias. Failure of fusion of the urethral folds results in maintenance of the secondary urethral groove. This condition is hypospadias. Concurrently, epispadias is occasionally present. In epispadias, which is due to failure in the development of the infraumbilical mesoderm, the dorsal wall of the urethra is partially or completely absent. In extensive cases, the bladder mucosa is exposed on the anterior abdominal wall (ectopia vesicae) together with cleft pelvis.

Classification

How are the types of hypospadia classified? Denis Browne[5] described five forms of this condition (Fig. 49). In descending order of frequency,[30] they are subcoronal (balanitic), penile, penoscrotal, scrotal, and perineal.

Complications

What are the most common complications in treatment of hypospadias? Most failures are the result of formation of fistulas, sloughing of skin, meatal retraction, recurrence of chordee, and infection.

Associated Malformations

What malformations are commonly associated with hypospadias? Chordee, cryptorchidism, urethral meatal stenosis, and the glans penis hooded by a redundant prepuce are the most frequently associated defects. Inguinal herina, anomalous upper urinary tract, and cardiovascular defects are sometimes encountered.

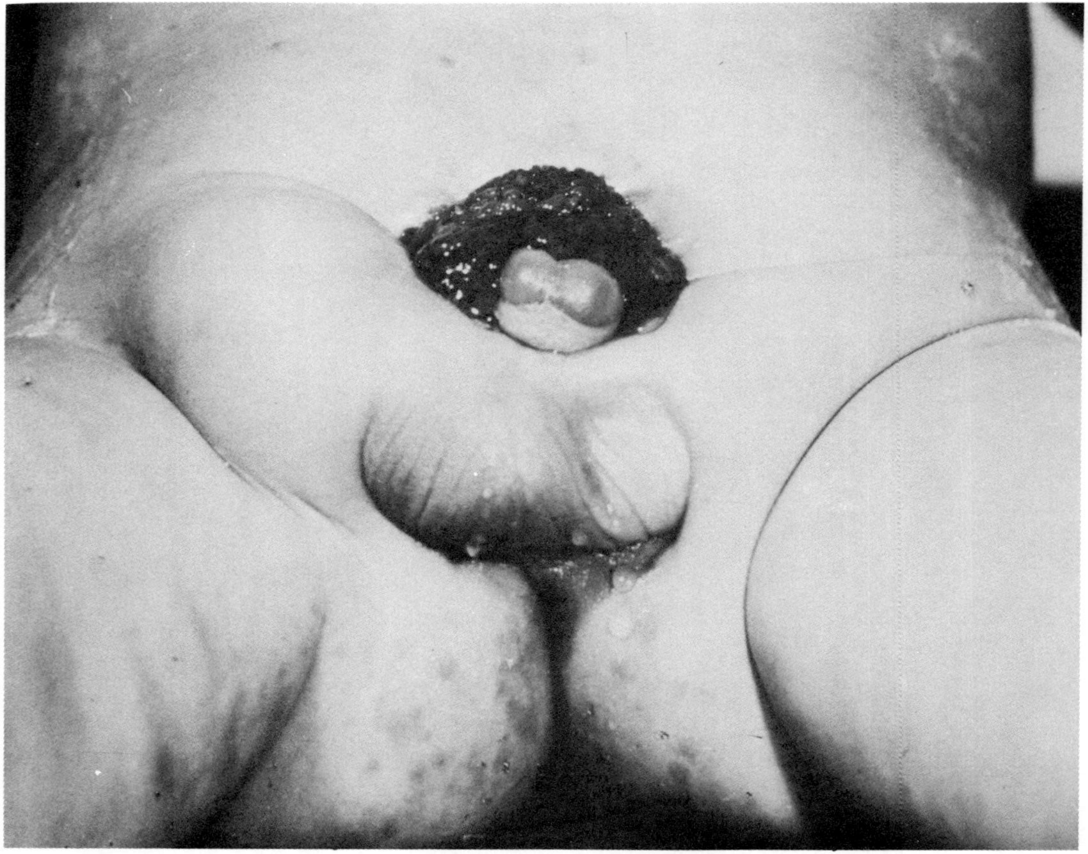

Exstrophy of the urinary bladder.

24

Exstrophy of the Urinary Bladder

A 7-day-old infant, product of a full-term normal delivery (3,300 gm.), was admitted to Children's Hospital for a defect of the anterior abdominal wall. At birth the anterior wall of both the urinary bladder and the urethra was lacking, thereby exposing the mucosa to the outside. On the lower part of the bladder mucosa were two small elevated projections (ureteral orifices) from which there was a continuous intermittent squirting of urine. When the child strained or cried, the intra-abdominal pressure and consequently the size of the bladder varied. Complete epispadias was present. Local contact with a dry diaper was painful, indicating an intact parasympathetic nervous system.

DIAGNOSIS

Exstrophy of the urinary bladder with complete epispadias (Grade II).

TREATMENT AND FURTHER COURSE

An excretory urogram was normal, and it was elected to readmit the infant at three-month intervals until the time of definitive repair. At home the exstrophied tissue was bathed frequently with water and exposed to warm, dry air and sunlight. The exposed mucosa was covered with a fine-mesh gauze coated with petrolatum to minimize the painful irritation from contact with the diaper and to retard hypertrophic and metasplastic changes.

Accompanying exstrophy of the urinary bladder were diastasis of the pubic symphysis, flared ilia, and posterolaterally rotated hip sockets which caused the infant to walk with a waddling gait.

At 18 months of age a definitive two-stage repair was done. The initial reconstruction was orthopedic. The patient was placed in a prone position, and after dissection of the soft tissue, the ilia were divided 2 cm. lateral and parallel to the sacroiliac joint to allow eventual coaptation of the pubic symphysis. After one week, the second stage of reconstruction entailing the bladder, urethra, pelvis, and abdominal wall was undertaken. The peritoneum was reflected from the bladder, allowing identification and mobilization of the right transverse band between the two halves of the pubic symphysis. This band, the external urethral sphincter, and its blood and nerve supply were carefully preserved. The bladder was catheterized and reconstructive repair of the urinary mucosa and sphincter was done around the indwelling catheter, after which the pubic symphysis was coapted. A plaster body cast with periodic changes was employed for six months. At 3 years of age the child had normal urinary control (continence with no reflux). The epispadias was repaired when the boy was 5 years of age prior to entrance into kindergarten.

Frequently, it is not feasible to reconstruct the bladder; the exstrophied mucosa is removed and only penile and urethral reconstruction is done. The urinary stream is then diverted by implantation of the ureters into the rectum or sigmoid colon, or by an ileal conduit.

DISCUSSION

Exstrophy of the urinary bladder occurs in 1 of 30,000 newborn infants.[30] Mackay and Syme[28] in 1849 advocated the application of external urinary receptacles in this malformation, and indicated little benefit from medical or surgical intervention. In the mid-1800's urinary diversion was attempted; Syme[43] performed the first successful ureterosigmoid anastomosis in 1852, but the patient died of ascending pyelonephritis. Since 1921, Coffey's[7] method of uretero-intestinal anastomosis has been improved and widely used. Reconstruction of the irregular bony pelvis that accompanies exstrophy of the bladder was first attempted in 1906 by Trendelenburg.[45]

Embryology

Discuss the cause(s) of exstrophy of the urinary bladder. The precise cause of exstrophy is not known, but the human embryo during normal development does not undergo a stage comparable to exstrophy or epispadias; therefore, the concept of developmental arrest cannot be applied. A widely accepted theory of the etiology of exstrophy was presented by Patten and Barry.[00] To understand the concept of Patten and Barry, it is necessary to briefly review the normal development of the cloacal region.

In the third week of development (Fig. 50A), there is a small region

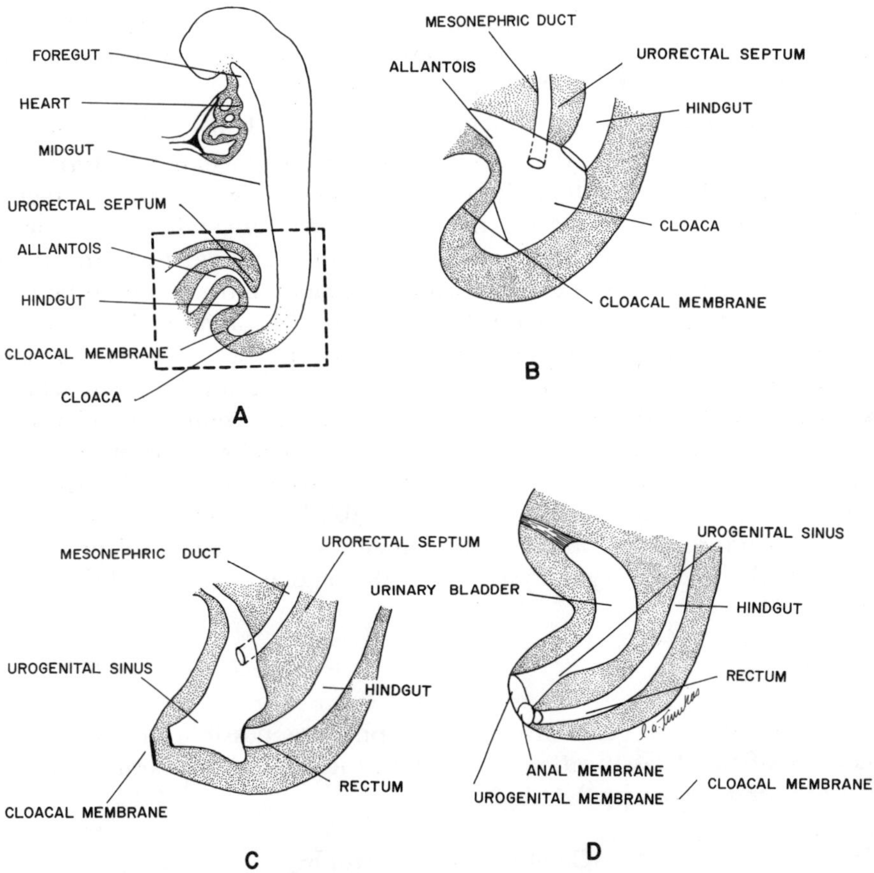

Fig. 50 Division of internal cloaca: *A*, 3-week-old embryo; *B, C*, and *D*, Enlargements of area outlined in *A*. *B*, 4-week-old embryo; *C*, 5- to 6-week-old embryo; *D*, 7-week-old embryo.

of contiguous ectoderm and entoderm in the midline of the embryo just caudal to the primitive streak; this is the cloacal membrane.

By the fourth week (Fig. 50B) the tail of the embryo has elongated and the cloacal membrane is just caudal to where the belly stalk joins the body. In this position the cloacal membrane now forms the ventral wall of the urogenital sinus. Mesoderm from the body wall on either side of the cloacal membrane migrates toward the midline, and by the fifth week (Fig. 50C) this mesoderm unites cranial to the cloacal membrane to form the genital tubercle. The mesodermal component of the infraumbilical portion of the ventral body wall converges toward the midline of the embryo, and as a result the distance between the umbilical cord and the cloacal membrane increases. At approximately the seventh week (Fig. 50D), the urorectal septum divides the cloacal membrane into a urogenital and an anal membrane.

The concept of Patten and Barry used to explain exstrophy of the urinary bladder suggests that the mesodermal components that form the genital tubercle arise lateral instead of cranial to where the urorectal septum fuses with the cloacal membrane; therefore, when the paired primordia of the genital tubercle fuse in the midline, the urogenital opening is cranial while the anal opening is caudal to the genital tubercle. The mesoderm of the genital tubercle, in this abnormal position—lateral to the urorectal septum—would not contribute to the formation of the infraumbilical portion of the abdominal wall, and in areas where ectoderm and entoderm make contact without intervening mesoderm, a rupture is imminent. When the urogenital membrane ruptures, the infraumbilical region of the body wall also splits, due to the lack of mesodermal reinforcement, and thus establishes exstrophy and epispadias. Rupture of the infraumbilical region of the body wall produces diastasis of the pubic symphysis, flared ilia, and posterolaterally rotated hip sockets.

Classification

What types of genitourinary exstrophy are possible and how are they classified? Classification is based on the clinical appearance of the bladder, urethra, and pelvic bones.[30]

Genitourinary Exstrophy

Grade I (minimal)—Diastasis of the pubic symphysis and rectus abdominus muscles is minimal. The urethra and external bladder sphincter are exstrophic.

Grade II (mild)—Diastasis of the pubic symphysis and rectus abdominus muscles is minimal. Exstrophy of the neck of the bladder and of the trigone expose the ureteral orifices. A considerable amount of bladder mucosa prolapses upon straining. The urethra is completely epispadic.

Grade III (moderately severe)—Grades I and II, and in addition there is bony involvement, i.e., diastasis of public symphysis, flared ala of the ilia, hips rotated posterolaterally, and feet pronated.

Grade IV (severe)—Grades I, II and III. In addition, there are extensive defects of the infraumbilical fascia, accompanied by one or more of the following: omphalocele, inguinal hernia, undescended testis, ventrally located anal orifice, relaxed anal sphincter, and intermittent rectal prolapse.

Associated Malformations

What malformations accompany exstrophy of the urinary bladder? This defect characteristically shows epispadias; inguinal, femoral, or umbilical hernias; and cryptorchidism. Additional abnormalities include vesicorectal fistula, rectal malformations, ureteral duplication, spina bifida, and hemivertebrae.

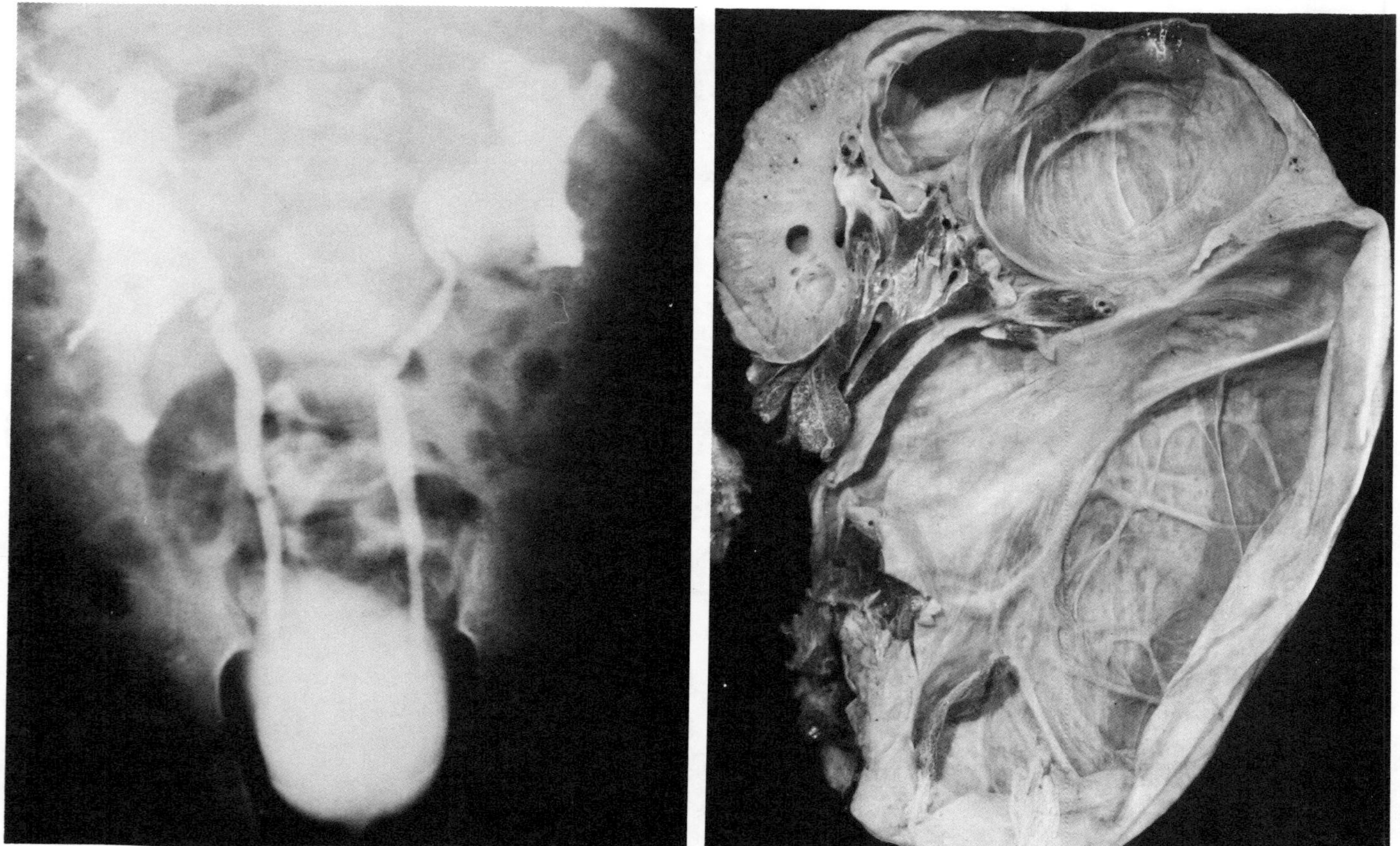

Left, Retrograde pyelogram of bilateral infantile polycystic kidneys; *Right,* Frontal section of a polycystic kidney.

25

Bilateral Polycystic Disease of the Kidney

A gravida I, para I mother delivered a full-term normal and well-developed male infant. When the infant was eight weeks of age, bilateral abdominal masses were palpated and an intravenous pyelogram was interpreted as polycystic kidneys. An open biopsy of the left kidney confirmed the tentative diagnosis. At four months of age the infant was admitted to the hospital and died shortly thereafter. Autopsy revealed an undernourished male infant (3,790 gm.) with deep eye furrows, snub nose, and micrognathia (Potter's facies). The abdomen was markedly distended due to the enlarged kidneys, the right kidney weighing 425 gm. and the left 235 gm. The external and cut surfaces of the kidneys showed innumerable thin-walled cysts that also were present in the cortex and the medulla. The ureters and bladder were normal. Additional cysts were seen in the portal zone of the liver.

Two years later the mother delivered a full-term female infant. Cyanosis was observed a few hours after birth. Physical examination revealed masses in both flanks, thought to be enlarged kidneys. The infant was placed in an oxygen tent and treated with antibiotics, but died 14 hours after birth. Postmortem examination showed a well-developed female infant (2,750 gm.) with facial features similar to those of her brother. The kidneys were enlarged, each weighing approximately 150 gm. The cut surfaces of the kidneys disclosed a large number of thin-walled cysts. The ureters and bladder appeared patent and normal. Microscopic examination demonstrated cysts in the kidneys and liver similar to those found in the brother. The mother has since given birth to two more children who are free of any renal disorder.

DIAGNOSIS

Bilateral infantile congenital cystic kidney.

DISCUSSION

Polycystic kidney disease has been recognized for several centuries, although its etiology has yet to be delineated. Since Kupffer[22] first demonstrated that the adult kidney develops from two separate anlagen, there has been much speculation as to how this process can result in polycystic kidney.[11, 20, 29, 31 32] Virchow[47] suggested that the structural changes within the kidneys were due to deposition of salts in the renal tubules, producing tubular obstruction followed by cyst formation. Numerous researchers[41, 44] attribute tubular obstruction to fetal syphilis.

Most investigators divide polycystic kidney disease into infantile and adult types,[16, 27] with some overlapping of their morphological features. Osathanondh and Potter (1964)[35] consider polycystic kidney to be a single disease which can be expressed as one of four types:

Type I Hyperplasia of the interstitial portion of the collecting tubules
Type II Inhibition of ampullary activity
Type III Multiple developmental abnormalities
Type IV Result of urethral obstruction

Bernstein (1970)[3] suggests that not all cystic abnormalities of the kidney are polycystic disease, and he classifies the types of cystic abnormalities as follows:

Bernstein's Classification of Cystic
Abnormalities of the Kidney (Modified)

- I. Polycystic disease
  - A. Infantile polycystic disease—genetic recessive
    1. Early infancy—corresponds to Osathanondh and Potter Type I
    2. Childhood
  - B. Adult polycystic disease—genetic dominant
- II. Cortical cysts
- III. Medullary sponge kidney
- IV. Hereditary and familial dysplasia
- V. Hereditary hydronephrosis and agenesis

The infants in the present case study correspond to Bernstein's Type I of polycystic disease; this is present at birth, always lethal, and is considered to be a structural defect that is autosomal recessive.

Cystic abnormalities of the kidney have also been classified according to: (1) genetic transmission; (2) pathology; (3) clinical course; and (4) roentgenographic presentation.[6, 16]

Embryology

Normal. Discuss the development of the normal adult kidney. Three distinct slightly overlapping kidney systems form during intrauterine life. They are, in order of appearance and increasing complexity: (1) the pronephros; (2) the mesonephros; and (3) the metanephros.

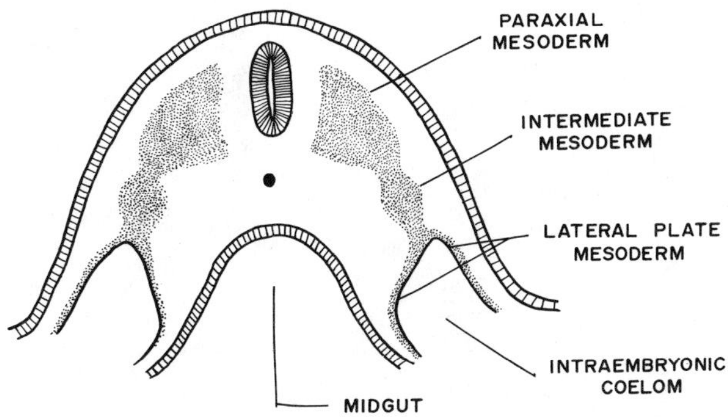

Fig. 51 Cross-section of a 2-week-old embryo showing division of intraembryonic mesoderm.

The intermediate portion of the intraembryonic mesoderm (Fig. 51) forms the nephrogenic ridge, which in the two-week-old human embryo differentiates into a series of segmentally arranged tubules, the pronephros. The caudal ends of these tubules join to form the pronephric duct, but by the end of the fourth week (Fig. 52) the pronephros itself and the cranial portion of the duct degenerate; the remainder persists to become part of the mesonephric duct.

The tubules of the mesonephros develop from the nephrogenic ridge caudal to the degenerating pronephros. These tubules acquire a glomerulus at their medial end, while at their caudal extremity they enter the remaining portion of the pronephric duct, now called the mesonephric (Wolffian) duct. The mesonephric duct grows caudually and enters the cloaca. Shortly prior to this the metanephric (permanent) kidney arises as the ureteric bud from the mesonephric duct close to its entrance into the cloaca (Fig. 52). The ureteric bud will eventually form the collecting portion of the metanephric kidney which consists of the collecting tubules, renal medulla, calyces, pelvis, and ureter.

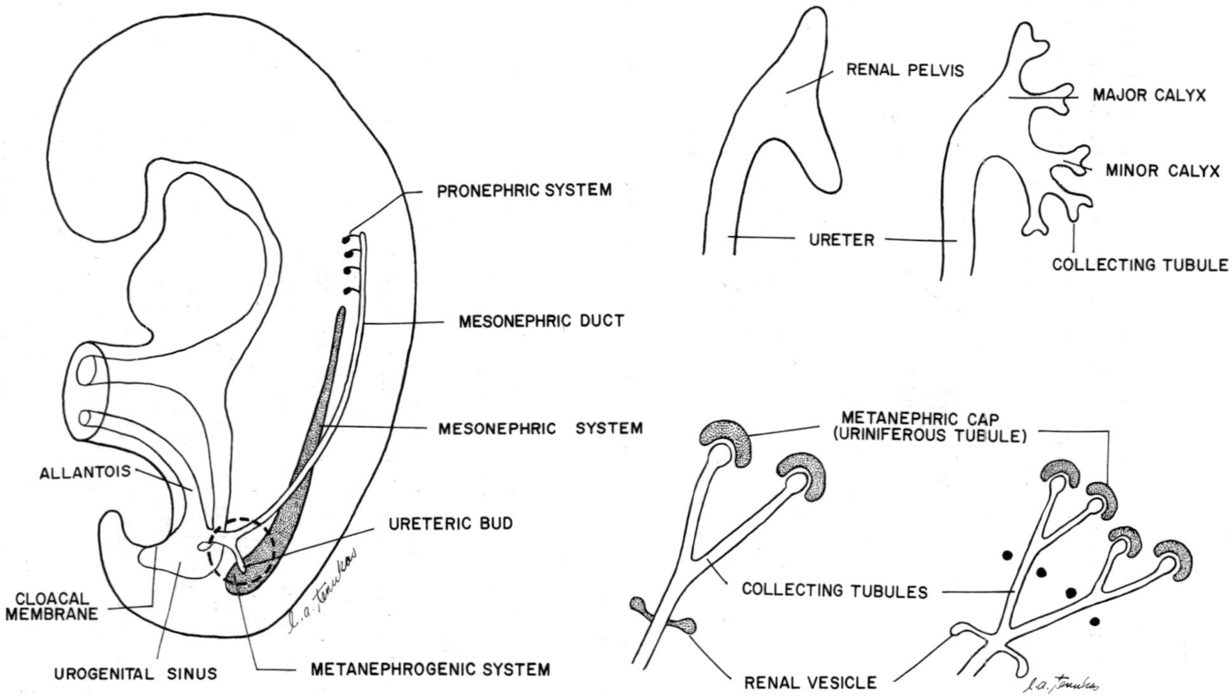

Fig. 52 Development of the normal kidney: 4-week-old embryo. *Right,* Enlargement of area circled: developing collecting system from ureteric bud.

The excretory portion of the metanephros (Fig. 52 right) develops from the intermediate mesoderm, which forms a mass of tissue called the metanephric blastema. The ureteric bud grows into the metanephric blastema; the former then expands, forming the major calyces. The major calyces continue to grow and subdivide through several generations of ducts, resulting in formation of the minor calyces and finally the definitive collecting tubules of the adult kidney.

The metanephric blastema forms a cap over the ureteric bud; portions of tissue separate from this cap and form clusters of cells, called renal vesicles, which become the excretory tubules or nephrons. The proximal end of the nephron forms Bowman's capsule while the distal end becomes confluent with the collecting tubule. Thus, a communication from the excretory to the collecting portion of the kidney is established.

Abnormal. Explain the formation of the infantile polycystic kidney. Hildebrand (1894)[18] proposed that failure of union between the excretory tubules of the metanephric blastema and the collecting tubules from the ureteric bud may result in a polycystic kidney. The accumulation of urine in the excretory convoluted tubules causes the nephron to dilate and eventually to form cysts. Kampmeier (1923)[20] recognized that the early generations of nephrons usually degenerate and disappear; cysts may be remnants of persisting nephrons. Subsequently several theories regarding cyst formation in the human kidney have been proposed, but are as yet unsubstantiated:

1. The collecting tubules do not unite perfectly with the excretory units, and as a result blind functional segments of excretory units are without a means of drainage into the collecting tubules.
2. Collecting tubules can develop an excessive number of branches, some of which may be incapable of inciting the development of metanephrogenic excretory tubules (nephrons); thus, some collecting tubules persist as blind ends.
3. The metanephrogenic intermediate mesoderm may show an incomplete response in tubular formation and arrest occurs before union to the collecting tubules is complete.
4. Nephrons become dilated (cystic) as a result of some obstruction to urinary outflow.

Associated Malformations

What malformations are closely associated with infantile polycystic kidney? Cysts of the liver, spleen, lungs, or pancreas, and esophageal atresia with tracheoesophageal fistula are most frequently associated with this defect. In addition, horseshoe kidney, anencephaly, and spina bifida are occasionally found with polycystic kidney.

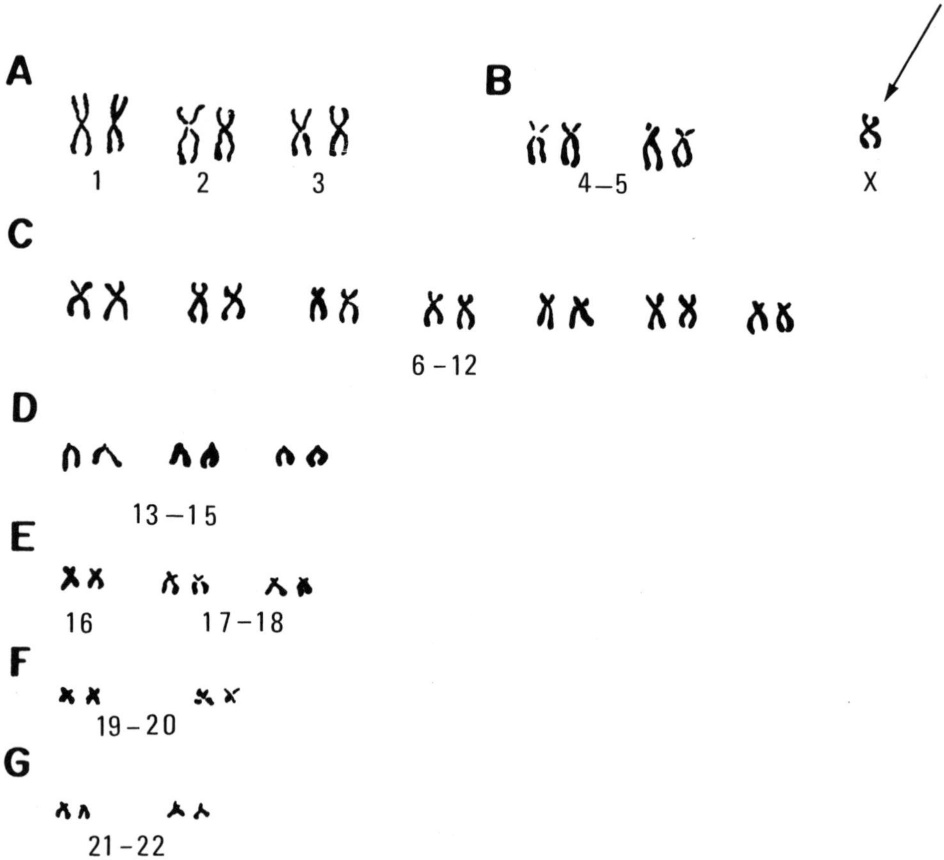

XO karyotype, *arrow,* of Turner's syndrome.

26

Gonadal Dysgenesis

The patient was the product of a full-term uneventful pregnancy and delivery in a 24-year-old primigravida mother. Physical findings suggestive of Turner's syndrome that were noted at birth included pterygium colli, edema of the hands and feet, hypoplasia of the nailbeds, and hypertelorism of the nipples. Walking, talking, and teething occurred at the normal ages.

At 16 years of age the patient saw a gynecologist because of primary amenorrhea, nonappearance of the secondary sex characteristics, and statural underdevelopment. Physical examination revealed retarded general somatic development. There was a lack of breast development, with no glandular tissue palpable and rudimentary nipples. The neck was short and webbed. There was a moderate cubitus valgus. Axillary hair was absent, pubic hair was scant, and the external genitalia were of infantile character; moreover, vaginal examination disclosed markedly underdeveloped labia. The vagina was too small for digital examination and the uterus and adnexa could not be palpated rectally.

The estrogen level was low. The urinary level of 17-ketosteroids was depressed, while the gonadotrophins were elevated. Vaginal smears showed atrophy of the epithelium. A buccal smear was negative for sex chromatin. The peripheral blood karyotype was 45X.

The patient received a cyclic dose of oral estrogen for two years, during which time the breasts developed and the external and internal genitalia matured. No menarche occurred. Treatment has continued.

DIAGNOSIS

Turner's syndrome.

DISCUSSION

What is gonadal dysgenesis and how is it classified? Gonadal dysgenesis is the absence of germ cells from the gonads (streak gonads) and the absence or underdevelopment of the female genital ducts and external genitalia. Gonadal dysgenesis can be categorized into three clinical types: (1) pure gonadal dysgenesis with no somatic anomalies; (2) gonadal dysgenesis with short stature as the only somatic anomaly; and (3) Turner's syndrome, a gonadal dysgenesis of the female accompanied by short stature, sexual infantilism, and at least two somatic anomalies.

The incidence of Turner's syndrome in patients with primary amenorrhea varies between 15 and 50 per cent.[40] In 1938, Turner[46] described a syndrome of sexual infantilism, congenital webbed neck, and cubitus valgus. Since Turner's description, many somatic anomalies have been associated with this syndrome. In 1956, Polani, Lessof, and Bishop[37] postulated that patients with Turner's syndrome have only one X chromosome; this was verified in 1959 by Ford and coworkers,[13] who published the first karyotype of Turner's syndrome.

Embryology

Discuss the development of the ovary. In the presomite embryo (3 weeks) (Fig. 53A), primordial germ (sex) cells are recognized within the yolk sac entoderm. The formation of the tail fold carries these germ cells into the wall of the hindgut where, via the dorsal mesentery, they migrate to the medial aspect of the mesonephric ridge to lie beneath the epithelium (Fig. 53B). The latter soon proliferates and bulges into the coelom as the genital ridge. Cellular (sex) cords enclosing the primordial germ cells develop in the genital ridge and penetrate the underlying mesenchyme. At this stage, the gonads are indifferent and, although the sex was determined at fertilization, histological differentiation is not yet possible. The female gonad differentiates (8-weeks' gestation) later than the male (6-weeks' gestation). The primordial germ cells (Fig. 53C, D) become surrounded by a capsule of flattened epithelial pregranulosa cells, thereby forming the primordial follicles which constitute the ovarian cortex. Mesenchyme then invades the gonad, separating the closely packed primordial follicles whose germ cells are now called oogonia. This mesenchyme is the ovarian stroma or medullary material. Many early follicles regress; moreover, secondary ingrowth of coelomic epithelium forms additional follicles. The period of proliferation of sex cords continues through the eighth month of gestation.

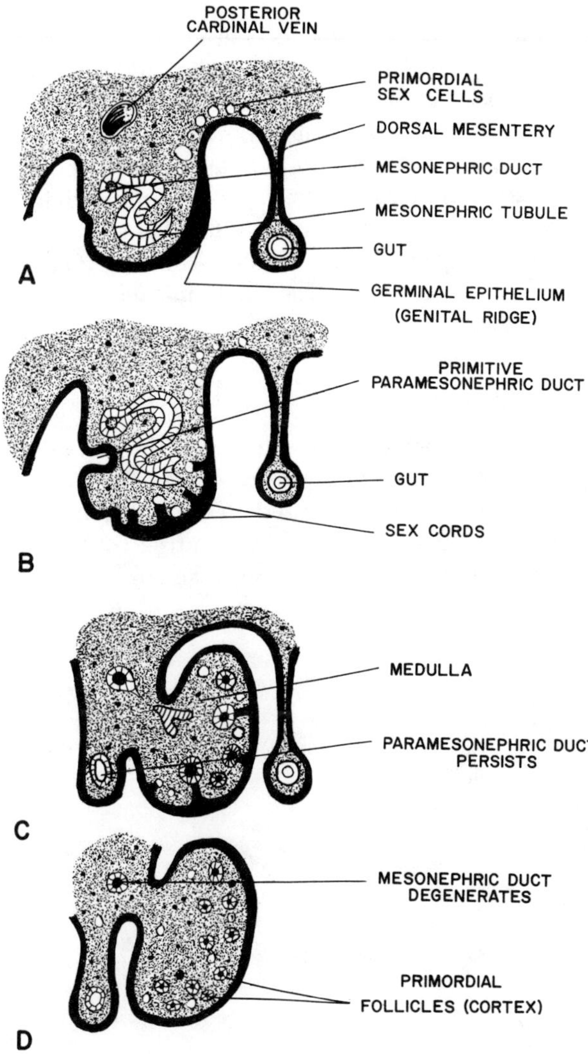

Fig. 53 Developing gonad in the female.

It is suggested that the migration of germ cells from the hindgut via the dorsal mesentery to the primitive genital ridge requires two active X chromosomes, and if one chromosome is lost as in Turner's syndrome or if part of it is deleted, migration does not occur and a streak gonad results.

Cytogenetic Findings

Explain the 45th X karyotype characteristic of Turner's snydrome. Numerical aberrations from the normal diploid number of chromosomes occur as two types: euploidy and aneuploidy. Euploidy deals with a balanced chromosome set in which the somatic complements are exact multiples of the basic number, while in aneuploidy the chromosome number is not the exact multiple of the basic set and individual chromosomes are missing or present in a multiple state. Thus, if one of two members of a pair of homologous chromosomes is missing (i.e., nondisjunction), the individual is monosomic for that particular chromosome as in Turner's syndrome.

Associated Malformations

Several malformations are associated with Turner's syndrome. Most commonly occurring are webbing of the neck, short neck, cubitus valgus, shield-like chest, multiple pigmented nevi, and hypoplastic nails. Other malformations include senile facies, epicanthic folds, high-arched palate, low-set ears, low hair line on the neck, coarctation of the aorta, and idiopathic hypertension.

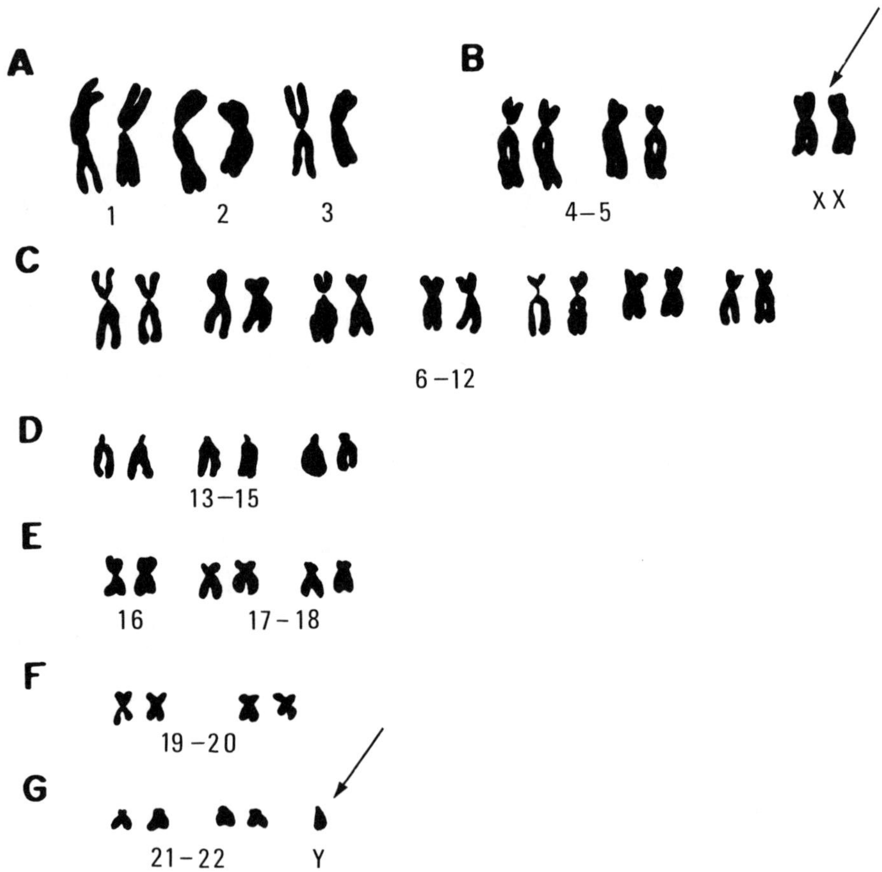

Karyotype of a Klinefelter patient with an XXY sex chromosome, *arrows,* complement.

27

Klinefelter's Syndrome

A 22-year-old married male was admitted to the hospital after complaining of sterility and gynecomastia. His general health had always been good. Puberty began at 14 years and progressed normally, although the testes remained small. At 16 years he first noticed an enlargement of the breasts which slowly increased until age 20. He had satisfactory erections, intercourse, and ejaculations.

Examination showed a tall, thin man who appeared younger than his stated age. Axillary hair was normal, pubic hair was of the female pattern, and the beard was scanty. The voice was rather high-pitched. The phallus was normally developed, while the prostate and scrotum were smaller than normal. The testes were extremely small.

Biopsy of the left testis showed hyalinization of the seminiferous tubules and relatively normal Leydig (interstitial) cells, accounting for the aspermia.

Laboratory studies revealed a high urinary excretion of gonadotrophin (FSH) and a deficiency of androgens. A buccal smear showed positive sex chromatin and a chromosome analysis disclosed a 47 XXY karyotype.

Gynecomastia was successfully treated by a subcutaneous mastectomy. There was no further treatment.

DIAGNOSIS

Klinefelter's syndrome (testicular dysgenesis, primary micro-orchidism, seminiferous tubule dysgenesis, true Klinefelter's syndrome, female pseudohermaphroditism with gonadal dysgenesis, chromatin-positive micro-orchidism, medullary gonadal dysgenesis).

DISCUSSION

The XXY karyotype of Klinefelter's syndrome occurs in 2.1 per 1,000 live male births.[8] Klinefelter, Reifenstein, and Albright (1942)[21] described a syndrome characterized by gynecomastia, aspermatogenesis without a-leydigism, and increased excretion of FSH. Polani and associates (1958)[38] suggested that these individuals had two X chromosomes, while the same year Ford and coworkers,[12] utilizing chromosome analyses, found an XX sex chromosome in bone marrow cells.

Classification

Depending upon somatic development, there are three types of Klinefelter's patients: (1) eunuchoid; (2) gonadal dysplasia; and (3) a type resembling the somatic norm.

Cytogenetic Findings

Aneuploidy refers to meiotic cell divisions, which result in an individual possessing an abnormal number of chromosomes; in Klinefelter's syndrome, meiotic cell divisions have produced an abnormal number of sex chromosomes. These accidents of cell division are of two types: (1) nondisjunction; two X's which should part, one into each derived nucleus, stay together and one nucleus gets two X's (Klinefelter's) while the other nucleus gets none (Turner's); and (2) anaphase lag in which the X and/or Y chromosomes fail to reach the spindle poles; hence their exclusion from the derived nuclei (Turner's).

REFERENCES

1. Amussat, J. Z.: Histoire d'une opération d'anus artificiel pratiquée avec succès par un nouveau procédé dans un cas d'absence congenitale de l'anus suivie de quelques reflexions sur l'obliteration du rectum. Gaz. Méd. Paris, Nov. 28, 1835.
2. Backus, L. H., and deFelice, C. A.: Hypospadias—then and now. Plast. Reconstr. Surg. 25:146, 1960.
3. Bernstein, J.: Morphology of inherited renal developmental abnormalities. Birth Defects Original Article Series 6:9, 1970.
4. Browne, D., and Sydney, M. B.: Some anatomical points in operation for undescended testicle. Lancet 224:460, 1933.
5. Browne, D.: An operation for hypospadias. Proc. Roy. Soc. Med. 42:466, 1949.

6. Carter, C. O.: Genetics of polycystic disease of kidney. Birth Defects Original Article Series 6:11, 1970.
7. Coffey, R. C.: Transplantation of the ureters into the large intestine in the absence of a functioning urinary bladder. Surg. Gynec. Obstet. 32:383, 1921.
8. Court Brown, W. M.: Sex chromosomes and the law. Lancet 2:508, 1962.
9. Dieffenbach, J. F.: Guérison des fentes congénitales de la verge, de l'hypospadias. Gaz. Hebd. Méd. 5:156, 1837.
10. Duplay, S.: De l'hypospadias perineo-scrotal et de son traitement chirurgical. Arch. Gén. Méd. 23:513, 1874.
11. Felix, W.: The development of the urinogenital organs. In Keibel, F., and Mall, F. P. (eds.): Manual of Human Embryology. J. B. Lippincott, Philadelphia, 1912.
12. Ford, C. E., Jacobs, P. A., and Lajtha, L. G.: Human somatic chromosomes. Nature (London) 181:1565, 1958.
13. Ford, C. E., Jones, K. W., Polani, P. E., de Almeida, J. C., and Briggs, J. H.: A sex-chromosome anomaly in a case of gonadal dysgenesis (Turner's syndrome). Lancet 1:711, 1959.
14. Frankl, O.: Einiges über die Involution des Scheidenfortsatzes und die Hüllen des Hodens. Arch. Anat. Entwcklngsgesch (Leipzig) p. 339, 1895.
15. Gregory, I. L.: Embryologic considerations in anorectal malformations. Dis. Colon Rectum 11:223, 1968.
16. Gwin, J. L., and Landing, B. H.: Cystic diseases of the kidneys in infants and children. Radiol. Clin. N. Am. 6:191, 1968.
17. Hepburn, R. H.: Anorchism. J. Urol. 62:65, 1949.
18. Hildebrand, O.: Weiterer Beitrag zur pathologischen Anatomie der Nierengeschwülste. Arch. Klin. Chir. 48:343, 1894.
19. Hunter, J.: A description of the situation of the testis in the foetus, with its descent into the scrotum. In Hunter, W.: Medical Commentaries. A. Millar, London, 1762.
20. Kampmeier, O. F.: A hitherto unrecognized mode of origin of congenital renal cysts. Surg. Gynec. Obstet. 36:208, 1923.
21. Klinefelter, H. F., Jr., Reifenstein, E. C., Jr. and Albright, F.: Syndrome characterized by gynecomastia, aspermatogenesis without a-leydigism, and increased excretion of follicle-stimulating hormone. J. Clin. Endocr. 2:615, 1942.
22. Kupffer, C.: Untersunchungen über die Entwicklung des Harn-und Geschlechtssystems. Arch. Mikr. Anat. 1:233, 1865.

23. Ladd, W. E., and Gross, R. E.: Congenital malformations of anus and rectum. Report of 162 cases. Am. J. Surg. 23:167, 1934.
24. Lambert, C. J.: Dartos pouch technique orchiopexy. Milit. Med. 132:690, 1967.
25. Littré, A.: Diverses observations anatomiques. Historie de l'academie royale de Science 23:36, 1710.
26. Lockwood, C. B.: The development and transition of the testis, normal and abnormal. J. Anat. Physiol. 22:505, 1888.
27. Lundin, P. M., and Olow, I.: Polycystic kidneys in newborns, infants and children. A clinical and pathological study. Acta Paediat. Upps. 50:185, 1961.
28. Mackay, J., and Syme, J.: Congenital exstrophy of the urinary bladder. Monthly J. Med. Sci. 9:934, 1849.
29. McKenna, C. M., and Kampmeier, O. F.: A consideration of the development of polycystic kidney. Trans. Am. Ass. Genitourin. Surg. 26:377, 1933.
30. Mustard, W. T., Ravitch, M. M., Snyder, W. H., Jr., Welch, K. J., and Benson, C. D. (eds.): Pediatric Surgery. Year Book Medical Publishers, Inc., Chicago, 1969.
31. von Mutach, A.: Beitrag zur Genese der congenitalen Cystennieren. Virchow. Arch. Path. Anat. 142:46, 1895.
32. Norris, R. F., and Herman, L.: The pathogenesis of polycystic kidneys. Reconstruction of cystic elements in four cases. J. Urol. 46:147, 1941.
33. Nové-Josserand, G.: Résultate éloignés de l'uretroplastie par la tunnellisation et la greffe dermo-épidermique dans les formes graves de l'hypospadias et de l'épispadias. J. Urol. Med. Chir. (Paris) 5:393, 1914.
34. Ombredanne, L.: Précis Clinique et Opératoire de Chirugie Infantile. Masson et Cie, Paris, 1923.
35. Osathanondh, V., and Potter, E. L.: Pathogenesis of polycystic kidneys. Historical survey. Arch. Path. 77:459, 1964.
36. Patten, B. M., and Barry, A.: The genesis of exstrophy of the bladder and epispadias. Am. J. Anat. 90:35, 1952.
37. Polani, P. E., Lessof, M. H., and Bishop, P. M. F.: Colour-blindness in "ovarian agenesis." Lancet 2:118, 1956.
38. Polani, P. E., Bishop, P. M. F., Lennox, B., Ferguson-Smith, M. A., Stewart, J. S. S., and Prader, A.: Colour vision studies and the x-chromosome constitution of patients with Klinefelter's syndrome. Nature 182:1092, 1958.
39. Rhodes, J. E., Pipes, R. L., and Randall, J. P.: A simultaneous

abdominal and perineal approach in operations for imperforate anus with atresia of the rectum and rectosigmoid. Ann. Surg. 127:552, 1948.

40. Rubin, A.: Handbook of Congenital Malformations. W. B. Saunders Company, Philadelphia, 1967.
41. Shukowsky, W. P., and Ssinjoff, W.: Ueber polycystische degeneration der Nieren namentlich im Kindesalter. Arch. Kinderheilk. 58:85, 1912.
42. Stephens, F. D.: Malformations of the anus. Aust. New Zeal. J. Surg. 23:9, 1953.
43. Syme, J.: Ectopia vesicae. Lancet 2:568, 1852.
44. Tarail, R.: Relation of abnormalities in concentration of serum potassium to electrocardiographic disturbances. Am. J. Med. 5:828, 1948.
45. Trendelenburg, F.: The treatment of ectopia vesicae. Ann. Surg. 44:281, 1906.
46. Turner, H. H.: A syndrome of infantilism, congenital webbed neck, and cubitus valgus. Endocrinology 23:566, 1938.
47. Virchow, R.: Ueber Hydrops renum cysticus congenitus. Virchow. Arch. Path. Anat. 46:506, 1869.

BACK, BODY WALL, AND LIMBS

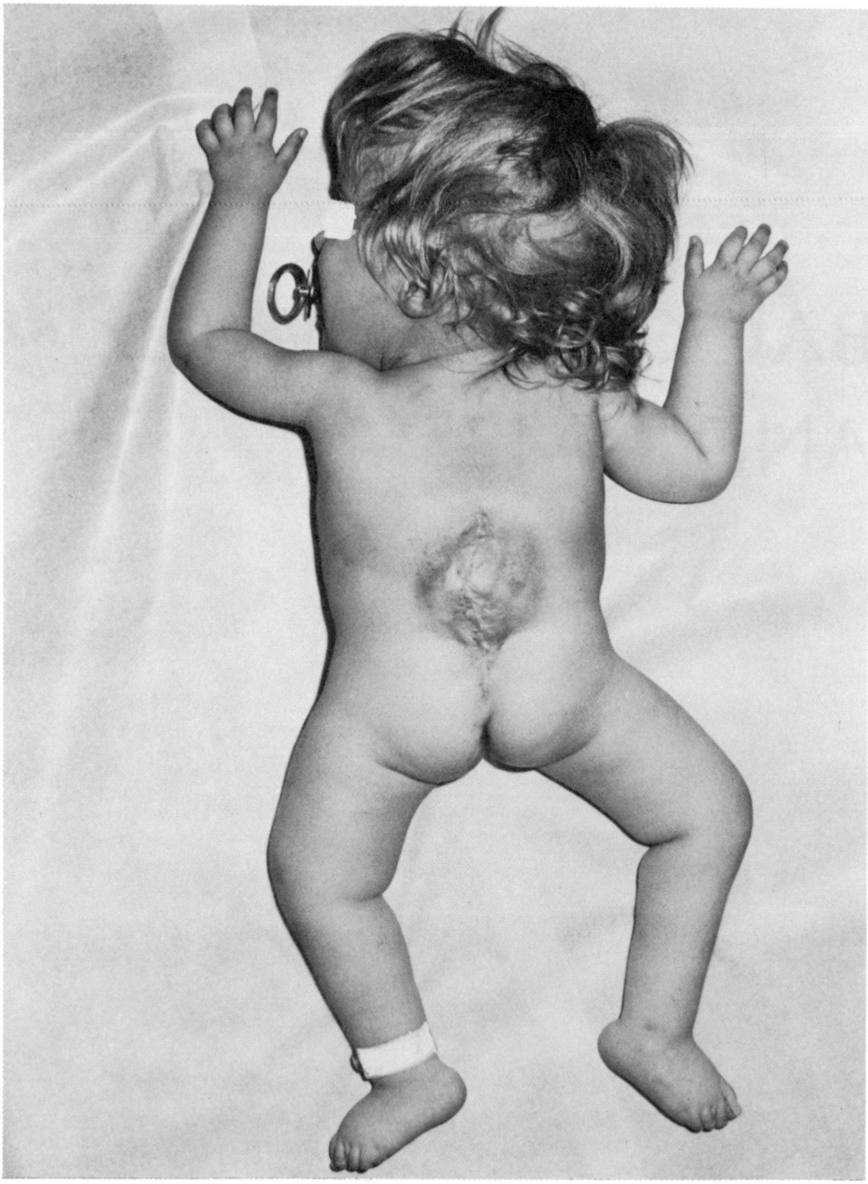

Repaired spina bifida cystica with myelomeningocele.

28

Spinal Dysraphia

A 35-year-old primiparous mother gave birth to a female infant. Moderate hydramnios was noted; otherwise the pregnancy and delivery were uneventful. Physical examination revealed a well-developed infant having a thin-walled, skin-covered sac in the lumbar region; the skin was thickened at the base of the sac. The left lower limb showed lack of spontaneous motion and sensory deficits in response to light pinprick. The anal sphincter revealed no loss of tone; the functional capacity of the urethral and vesical sphincters could not be ascertained at that time. Radiographic examination indicated a fusiform dilatation of the spinal canal centered at the site of the lumbar sac. In addition, there was an increased distance between the thinned and tilted vertebral pedicles. Neural elements were identified in the lumbar sac using transillumination. A left talipes equinovarus deformity was present. The family history revealed spinal dysraphia and clubfoot defects.

DIAGNOSIS

Spina bifida cystica posterior (spina bifida manifesta, spina bifida aperta, SBC) with myelomeningocele (Fig. 54B) and left clubfoot.

TREATMENT

At 24 hours of age the infant was prepared for surgical excision of the myelomeningocele. The superficial tissue between the membranous neck of the myelomeningocele sac and the surrounding fascia and muscle was separated. The superior aspect of the sac was opened to avoid injuring the neural contents, and with release of the cerebrospinal fluid the skin and membranous covering relaxed. The neural elements were dissected free from the myelomeningocele sac and allowed to drop into the spinal canal, and the sac was excised. A watertight closure of the dura and skin was completed. Postoperatively, the infant was placed in a prone position and watched closely for evidence of increased intracranial pressure: hydrocephalus as a result of possible abnormal circulation of the cerebrospinal fluid. At six weeks of age hydrocephalus became evident and was relieved employing a ventriculoatrial shunt. Also present was a loss of tone of the urethral and vesical sphincters, which resulted in dribbling of urine. Urinary tract function and possible damage by urinary stasis were evaluated using regular intravenous pyelograms and cystograms. Evidence of upper urinary tract impairment occurred at five years of age, and a urinary diversion, ureteroileostomy, arrested further damage. Orthopedic care of the paralyzed left lower limb and clubfoot was continued; further growth and development were normal.

DISCUSSION

Spina bifida occurs in 1 out of 1,000 births in the United States.[31] Nicolaas Tulp (1641)[38] introduced this name and gave the first precise description of the lesion. Tulp described three cases having cystic lesions and provided two plates which clearly show the gross abnormalities. Morgagni[21] recognized the association of spina bifida malformation with abnormalities such as hydromyelia and hydrocephalus, and believed the bifid spine occurred as an erosion by cerebrospinal fluid under increased pressure. Cruveilhier (1832)[38] attributed spina bifida to developmental errors. The syndrome of spina bifida occulta was defined and named by Virchow in 1875.[22] The various forms of spina bifida cystica are classified according to the pathologic studies of von Recklinghausen (1886).[29]

Embryology

Discuss the early formation of the spinal cord and its skeletal covering. The neural tube (Fig. 16) develops from thickened surface ectoderm, the neural plate, at three-weeks' gestation. Soon the meninges and vertebral column (Fig. 17) begin to form. The neural plate overlies the notochordal process and the medial edge of the paraxial mesoderm, and becomes depressed axially to form the shallow neural groove. In early somite embryos, in the future cervical region the lateral margins of the neural groove fuse, forming the neural tube. The primordium of the future brain is immediately cranial to the cervical region, while the spinal cord is caudal. By the fourth week the neural tube has closed except at the cranial (anterior) and caudal (posterior) neuropores; closure here occurs at 24 and 26 days, respectively. As the somite stage ends, the spinal cord is a closed tube that extends the length of the embryo and is surrounded by the meninges. The cartilage and bone composing the vertebral column grow more rapidly than the cord; thus, by full term the coccygeal end of the cord lies at the level of the third lumbar vertebra.

The skeletal covering of the spinal cord is the vertebral column, which arises in the fourth week of gestation from the sclerotomic mesenchyme of the axial mesoderm. The sclerotomes occur in pairs on either side of the notochord, each pair separated from another by an intersegmental artery. By the sixth week each sclerotome is divided into a cranial portion and a more dense caudal portion. The caudal portion unites with the cranial portion of the sclerotome just caudal to it, thereby forming the primordia of the definitive vertebra. The sclerotomes then grow in three directions: (1) mesad—to surround the notochord and form the vertebral body; (2) dorsad—to flank the neural tube and form the vertebral arch; and (3) ventrolaterad—to form the costal processes in the interspaces between the epaxial portions of the myotome. Mesenchymal tissue from two abutting vertebrae condenses to form the intervertebral disk; remnants of the notochord are incorporated into the disk and persist as the nucleus pulposus.

Etiology

What factors may cause spina bifida cystica posterior? Spina bifida cystica, a defect of the bony encasement of the spinal cord through which a protrusion of meninges, cord, or both occur, may result from: (1) a noxious chemical stimulus affecting the organizer and resulting in neural tube nonclosure or neural tube rupture and (2) environmental factors, e.g., hypervitaminosis A, salicylate poisoning or X-radiation. Advanced maternal age also has been implicated.

Classification

Distinguish between the various forms of spina bifida. Classification of spina bifida is dependent upon the degree of involvement of the vertebrae and/or the spinal cord.

1. Spina bifida anterior — a defect of the anterior surface of the bony spinal canal associated with de-

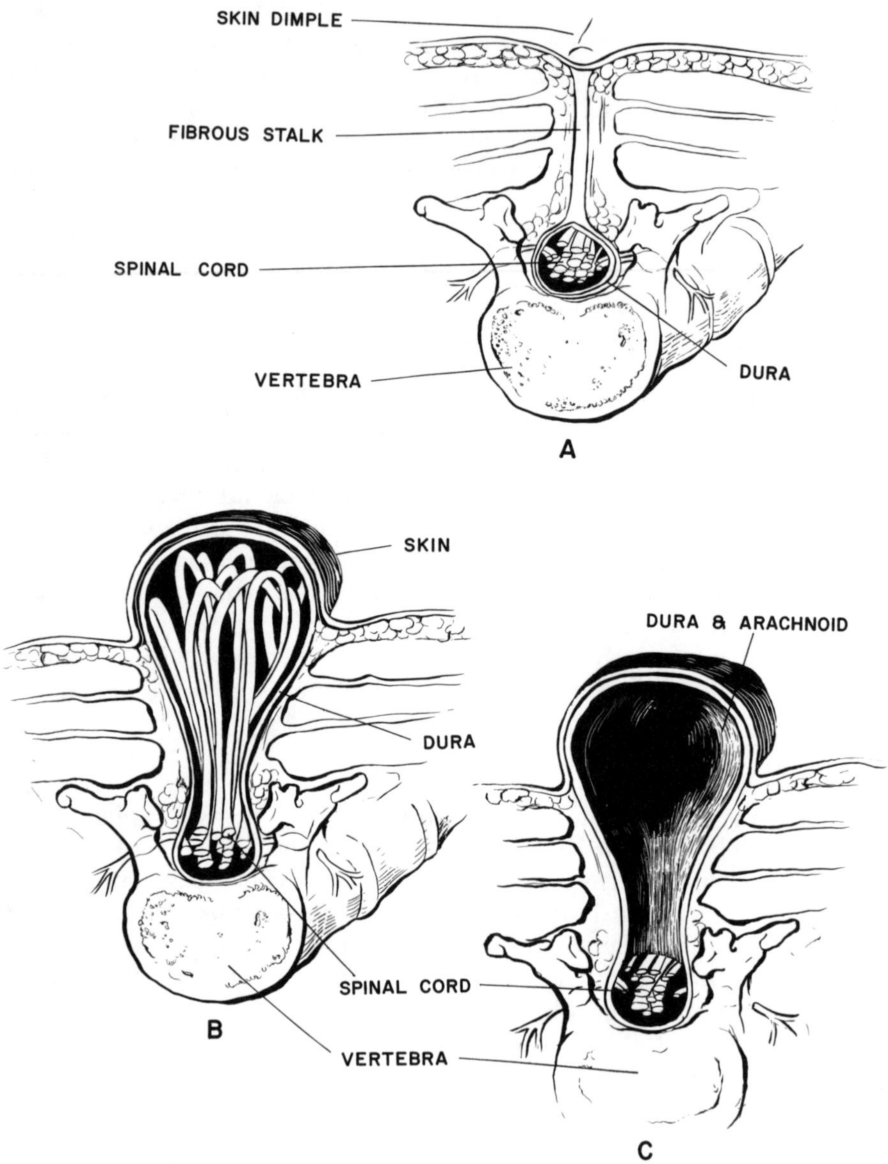

Fig. 54 Spina bifida: *A*, Spina bifida occulta; *B*, Spina bifida cystica with myelomeningocele; *C*, Spina bifida cystica with meningocele.

		fective development of the thoracic and abdominal viscera.
2.	Spina bifida posterior	– a defect of the posterior surface of the bony spinal canal.
3.	Spina bifida occulta (Fig. 54A)	– a defect of the bony spinal canal without protrusion of the cord or meninges.
4.	Spina bifida cystica	– a defect of the bony spinal canal with a protrusion of a cystic swelling involving:
	a. myelomeningocele	– meninges plus spinal cord (Fig. 54B)
	b. meningocele	– meninges (Fig. 54C).

Associated Malformations

What malformations may occur with spina bifida cystica? Hydrocephalus, aqueductal stenosis, motor and sensory disturbances, clubfoot, anomalies of the urinary tract, congenital dislocation of the hip, and anencephaly are the most common defects accompaning spina bifida cystica.

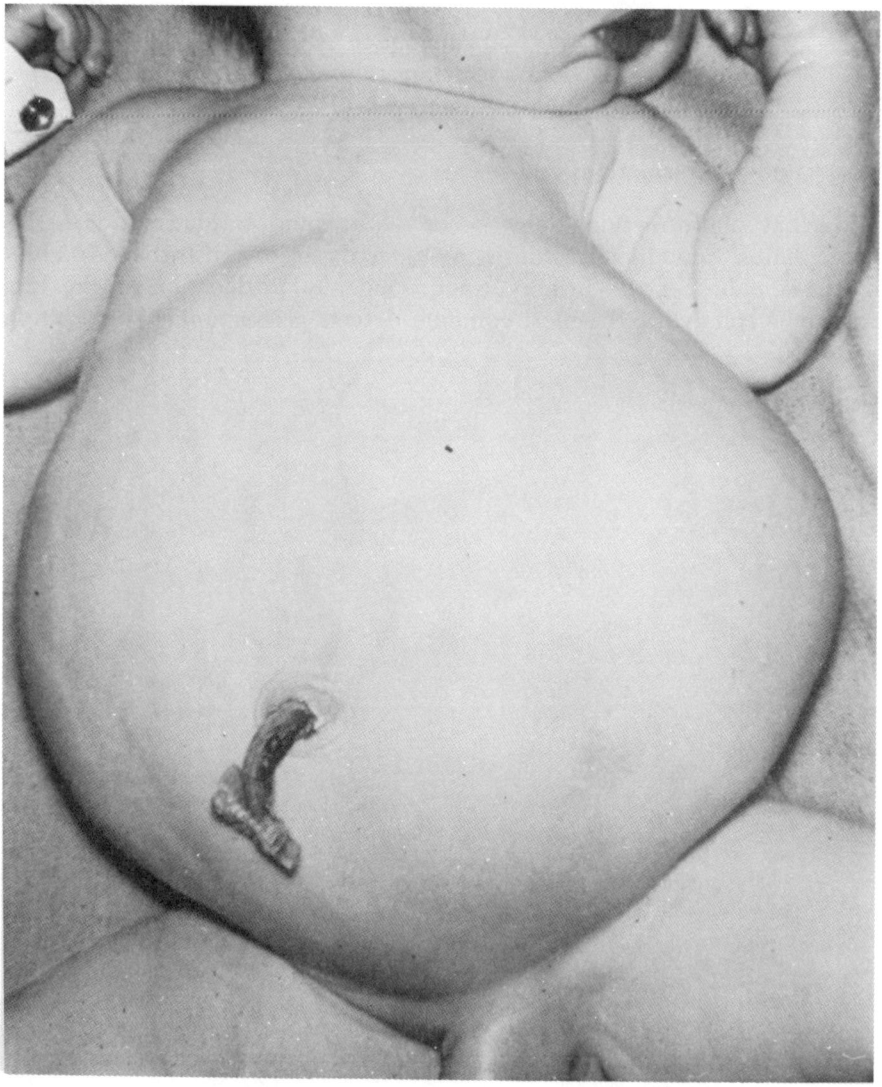

Congenital absence of abdominal musculature (prune belly).

29

Abdominal Musculature Deficiency Syndrome (AMD)

A full-term male infant was noted at birth to have an enlarged, flaccid abdomen. Examination revealed the absence of abdominal musculature, causing the skin to appear overabundant and wrinkled with the abdomen bulging into the flanks. The abdominal organs, especially the kidneys and the urinary bladder, were easily palpable. The intestinal loops and persistalsis were visible. The gastrointestinal and genitourinary tracts were explored for malformations that commonly accompany the absence of abdominal musculature. An abdominal binder was used to protect existing muscle elements and to prevent overstretching. The testes were undescended, but the scrotum was sufficiently developed.

DIAGNOSIS

Congenital absence of the abdominal musculature (prune belly,[39] triad,[23] or Eagle-Barrett syndrome[6]).

TREATMENT

At six months of age the infant demonstrated painful micturition and was admitted to Children's Hospital. Urography showed poor bilateral kidney function and nephrostomies were performed. Subsequently no drainage occurred from the left side. The abdomen was opened; the left ureter was dilated and an atretic left kidney having small cysts was located. The kidney was removed. The right ureter that was markedly dilated and tortuous was mobilized and reimplanted into the bladder, which was enlarged and attached at its apex to the umbilicus

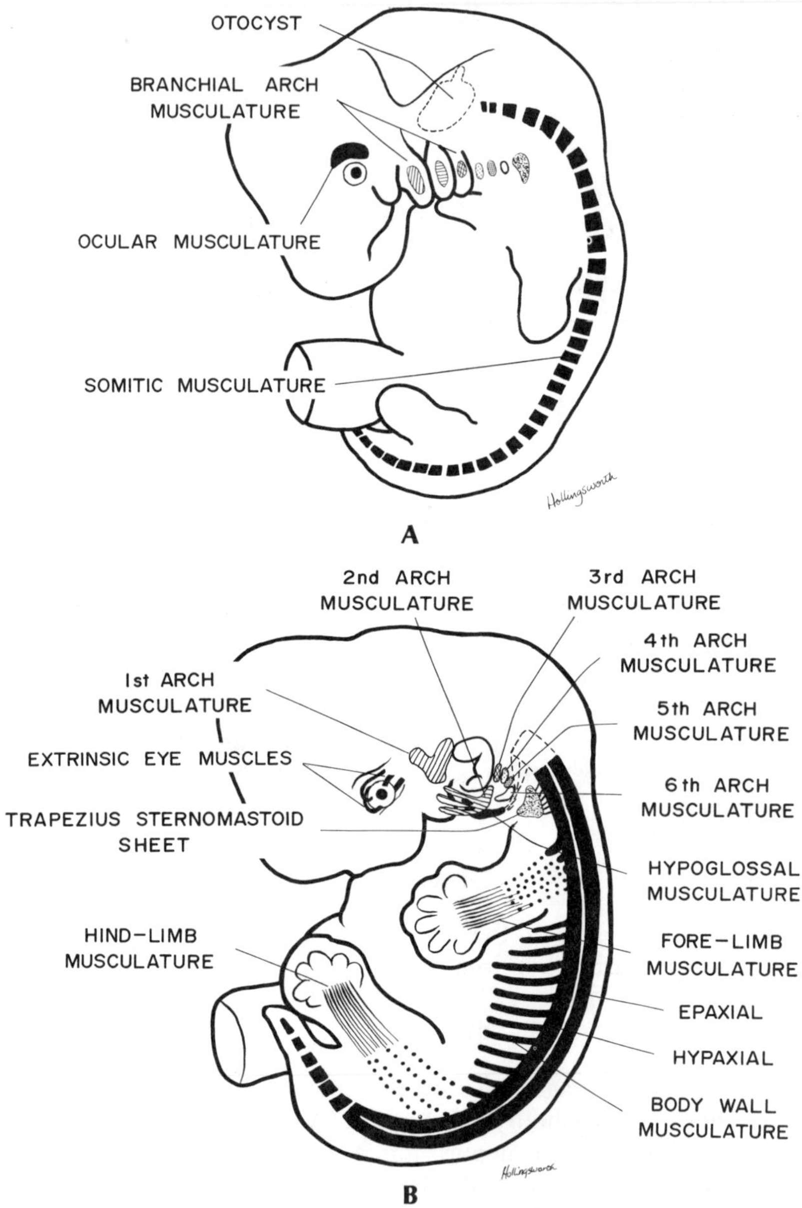

Fig. 55 Sequential development of muscle masses: *A,* 3.5-week-old embryo; *B,* 5-week-old embryo; *C,* 3-month-old fetus. (From Hamilton, W. J., Boyd, J. D., and Mossman, H. W.: Human Embryology, ed. 3. W. Heffer & Sons, Cambridge, England, 1964, with permission.)

by the urachal diverticulum. When the infant was ten months old, a micturating cystogram demonstrated an enlarged bladder. There was no reflux or urethral obstruction and the prostatic urethra showed filling of a prominent prostatic utricle. After the urinary infection subsided, the patient was remarkably symptom free and at five years of age showed little change. The abdominal wall had strengthened with an improved appearance. The undescended testes were treated at a later date.

DISCUSSION

The earliest reports of congenital absence of the abdominal muculature were by Fröhlich (1839),[10] von Ammon (1842)[1] and Henderson (1890).[13] There are less than 200 reported cases, with the majority occurring in the male. Parker (1895)[26] noted that anomalies of the urinary tract accompany this malformation and form part of a syndrome.

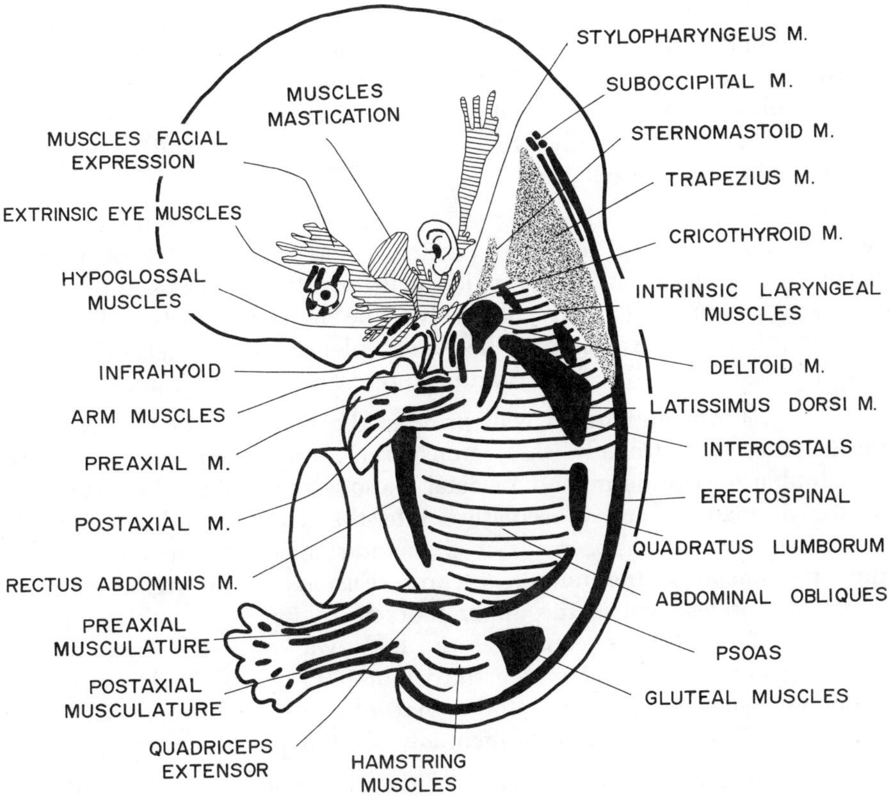

C

Fig. 55 (continued. See facing page for legend.)

Embryology

Describe the muscle formation and closure of the anterior abdominal wall. Following establishment of the trilaminar embryo early in the third week of gestation, the intraembryonic mesoderm on either side of the notochord thickens to form the paraxial mesoderm. By the twenty-first day (Fig. 55A) this paraxial mesoderm begins to segment into paired cubical masses, the somites. The first somite (occipital) forms at the cephalic end of the notochord; successive somites progressively differentiate. No somites form in the mesoderm cranial to the notochord; however, paraxial mesoderm, as diffuse mesenchyme, extends to the cephalic end of the future brain plate. Totally, 42 to 44 pairs of somites form (4 occipital, 8 cervical, 12 thoracic, 5 lumbar, 5 sacral, and 5 to 10 coccygeal).

The established somite differentiates into three components: (1) the myotome, (2) the dermatome, and (3) the sclerotome. The dermatome migrates under the surface ectoderm to form the dermis and subcutaneous tissue of the skin, while the sclerotome becomes associated with the notochordal process. The myotome enlarges dorsally (to flank the neural tube) and ventrally (to extend into the body wall) and gives rise to voluntary musculature.

In the fifth to sixth week (Fig. 55B), the myotome divides into two unequal portions, a dorsal portion (epaxial) and a ventrolateral portion (hypaxial) (Table 3), which will form the extensors and flexors of the vertebral column, respectively. The hypaxial portion loses its segmentation and grows into the somatopleure as a sheet of muscle which gives rise to the ventral and lateral trunk musculature.

The anterior abdominal wall prior to the migration of the hypaxial musculature is extremely thin, transparent, and composed solely of somatopleure (ectoderm and parietal mesoderm). With the extension of the dermatome and musculature into the anterior abdominal wall, the transparent area becomes a diamond-shaped region surrounding the attachment of the umbilical cord. Fusion of the right and left hypaxial muscles, first in the upper thoracic region, suprapubically, and finally at the umbilicus, completes the body wall with the formation of the linea alba at the end of the third month (Fig. 55C). The external abdominal oblique, internal abdominal oblique, and transversus abdominis muscles form from tangentially split hypaxial musculature.

Failure of the hypaxial portion of the myotome to migrate ventrally or to complete its growth results in absence or partial development of the anterior abdominal wall musculature, with the abdominal wall retaining the thin transparent appearance characteristic of the early embryo.

Etiology

What etiological factors may contribute to the prune belly syndrome (absence of abdominal wall musculature and abnormalities of the genitourinary tract)? The etiology of this syndrome remains obscure; nevertheless, there are three main speculations, none of which are totally acceptable:

1. Stumme[35] suggested that the primary defect is an outlet (urethral) obstruction leading to bladder enlargement, which results in sufficient pressure upon the abdominal musculature to cause its atrophy.
2. Osler[25] stated that the primary defect might be the aplastic abdominal musculature, which offers no resistance to an obstructed, expanding urinary bladder.
3. Silverman and Huang[35] offered the most acceptable explanation. Between the fifth and tenth weeks of development, the abdominal wall and the gastrointestinal and genitourinary tracts are in a stage of development such that an insult can result in malformation. Such malformations are associated with one another only by the proximity of time and place, and are not otherwise interrelated.

Associated Malformations

What abnormalities are most frequently associated with congenital absence of the abdominal musculature? Gastrointestinal and genitourinary tract abnormalities most commonly accompany this malformation. Clubfoot, congenital dislocation of the hip, sternal and thoracic defects, and polydactylism may be present.

Table 3. Origin of the segmental musculature*

Myotomes	Derivatives	Nerves
Epaxial Cervical Thoracic Lumbar	extensors of the vertebral column	$C_1 - S_{3\;4}$
Sacral Coccygeal	dorsal sacral ligaments formed by degenerative or retrogressive changes	
Hypaxial Cervical	scalenes, prevertebrals, geniohyoid, and infrahyoid	$C_1 - C_8$
Thoracic and lumbar Anterior and lateral thoracic and abdominal wall	intercostals, sternalis, rectus abdominis, external abdominal oblique, transversus abdominis	$T_1 - L_1$
Posterior abdominal wall	quadratus lumborum	$L_1 - L_5$
Sacral and coccygeal (?) Pelvic diaphragm	coccygeus and levator ani	$S_2 - Co_1$
Cloacal	anal sphincter and external genitalia musculature	$S_2 - Co_1$

*Areas where direct evidence is disputed: (1) orbital (extrinsic eye muscles) CN III, IV, VI; (2) tongue CN XII; and (3) limbs; upper $C_4 - T_1$, lower $L_2 - S_3$. Musculature of these areas probably develops from in situ mesenchyme.

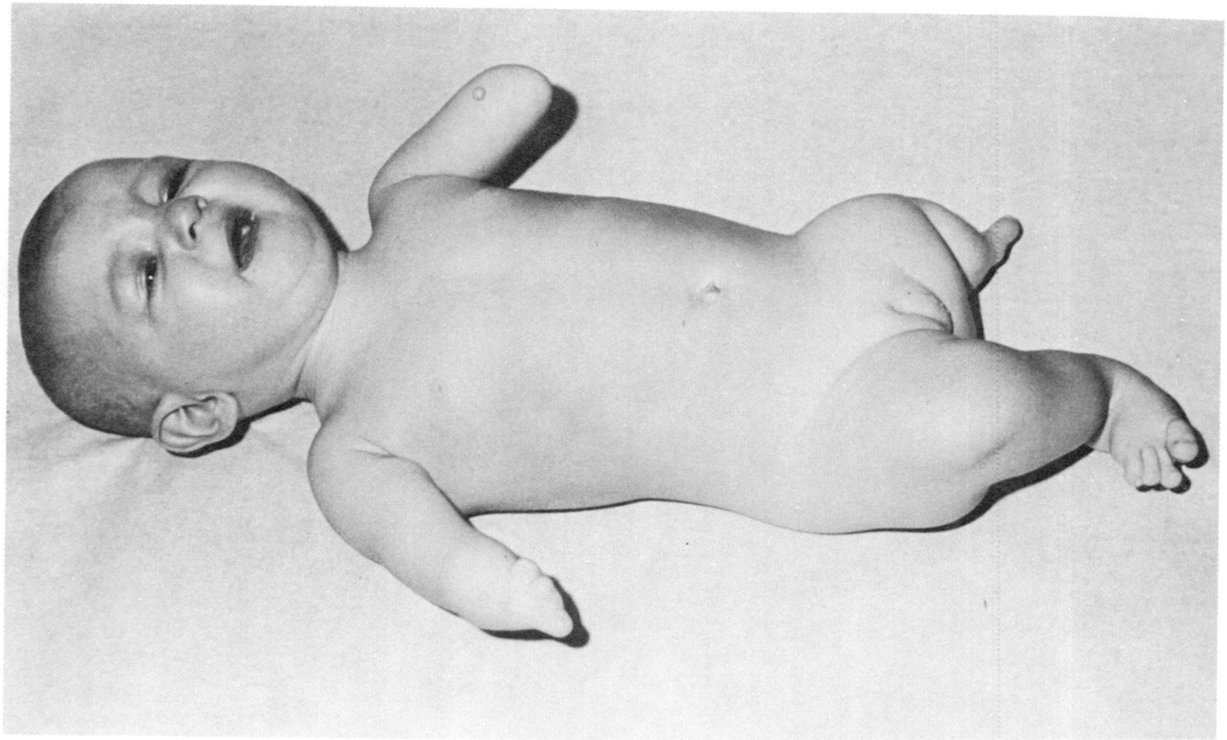

Dysmelia (etiology unknown).

30

Thalidomide Syndrome

In 1961 an Australian 25-year-old primipara female, having a history of 37 weeks amenorrhea, was in labor and admitted to the hospital. Four hours later she delivered a female infant. Initial examination of the infant showed the upper limbs to be malformed; the hands were set directly into the shoulder region, both appearing like flippers. Hemangioma of the forehead, nose, and upper lip was present as well as absence of the left external ear with the internal auditory canal situated abnormally low in the head. The trunk and lower limbs appeared normal. Radiological examination of the upper limbs revealed absence of the long bones, with three ulnar finger rays (minimus, annularis, medius) and ulnar parts of the carpus attached to a hypoplastic shoulder girdle. No additional malformations were noted. The mother's history revealed that during the fourth week of pregnancy vomiting occurred and the drug Distaval was administered as an antiemetic up to the eighth week of gestation.

DIAGNOSIS

Thalidomide syndrome (bilateral intercalary phocomelia, facial hemangioma, and auditory malformation). Abnormalities involving the long bones of the upper limbs characterize the majority of cases, with lower limb involvement in approximately half of these infants. In addition, a saddle-shaped or flattened nose, unilateral paralysis of the face, and malformations of the internal organs are seen.

FURTHER COURSE

The child appeared to have normal intelligence and, in spite of the ear deformity, hearing was normal. Prosthetic fittings were delayed until the functional capabilities of the phocomelic limbs could be assessed.

DISCUSSION

Prior to the introduction of thalidomide, alpha(N-phthalimido)—glutarimide, few cases of phocomelia (1/75,000[31]) or other malformations of the long bones were reported. Pliny[27] wrote of a race of men with one median lower limb, and Batholin[2] showed this abnormality in his Anatomic Atlas. Several paintings depict limb malformation; Handschriften[11] portrays an armless town clerk of Nuremberg during the sixteenth century, and Goya,[34] in a painting now in the Louvre, shows limb malformation. The classification of limb anomalies by Saint-Hilaire in 1832[32] is still in great part used today.

A Swiss pharmaceutical house first synthesized thalidomide in 1954. The drug showed no effects in laboratory animals and the company discarded it. Then the West German firm, Chemie Grünenthal, pursued the development of thalidomide and used it as an anticonvulsant for epileptics, a sedative for mental patients, and an antiemetic and sedative for pregnant women. In 1961, Lenz[16] in West Germany attributed the rapid increase in limb malformations in newborn infants to Contergan, while that same year McBride[19] in Australia suggested Distaval as the cause. Both drugs were trade names for thalidomide. Unfortunately, thalidomide was sold under several names and time elapsed before all preparations were removed from the market.

Table 4. Thalidomide preparations

Trade Name	*Location*
Distaval	British Isles Australia New Zealand
Softenon	Portugal
Talimol Kevadon	Canada
Contergan	West Germany
Kevadon	United States—never released for sale by F.D.A.

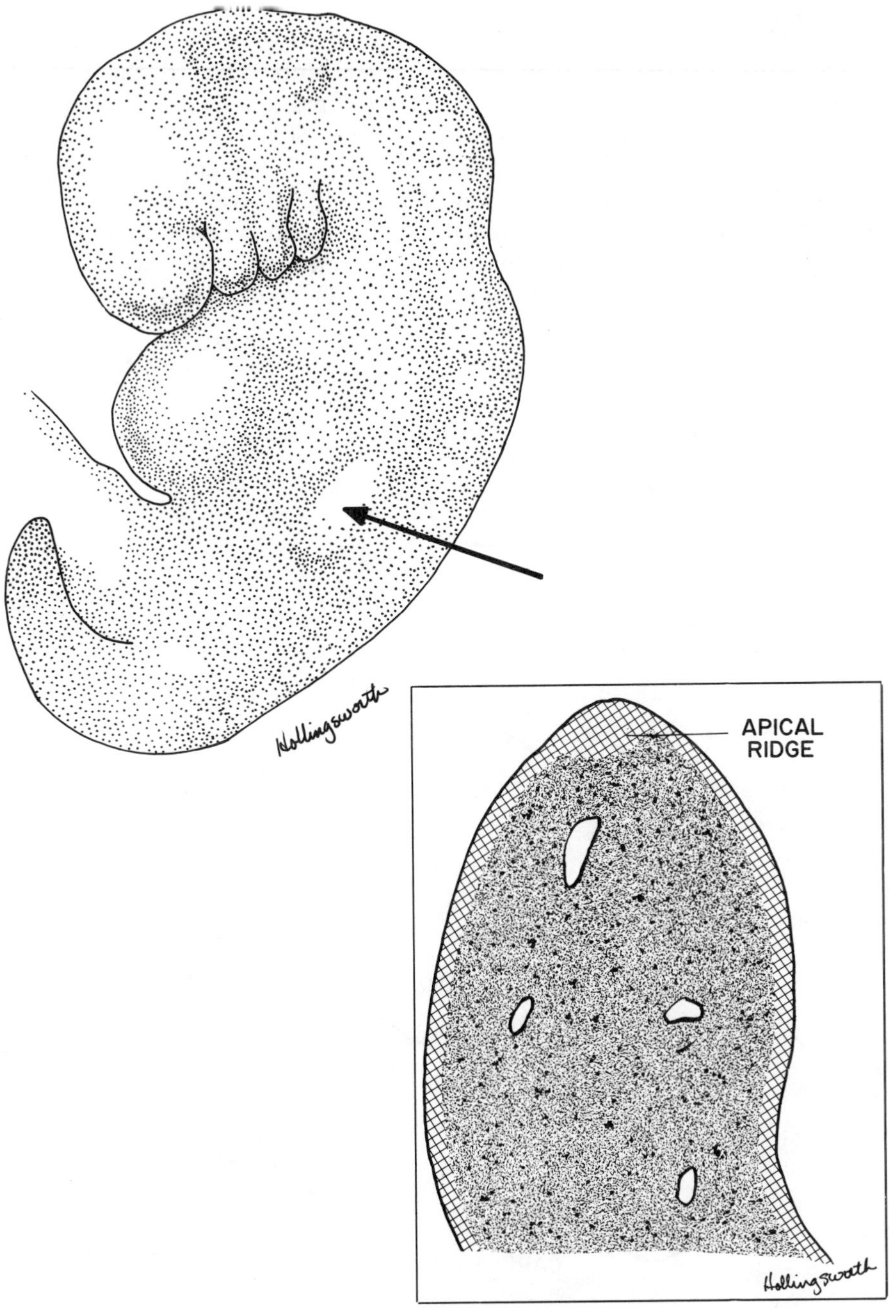

Fig. 56. Upper limb development, *arrow*, in 3.5-week-old embryo. *Inset:* Apical ridge.

Discuss the development of the upper limb. Discuss the possible causes of phocomelia. Upper limb buds first appear as longitudinal mesodermal ridges, which induce a thickening of the overlying ectoderm (apical) on the body wall between C_2 and T_1 on approximately the twenty-sixth day of gestation (Fig. 56, inset). Lower limb buds soon appear and are markedly retarded in development when compared with the upper limbs. The upper limb buds become rounded, then flattened dorsoventrally, acquiring cephalic (preaxial) and caudal (postaxial)

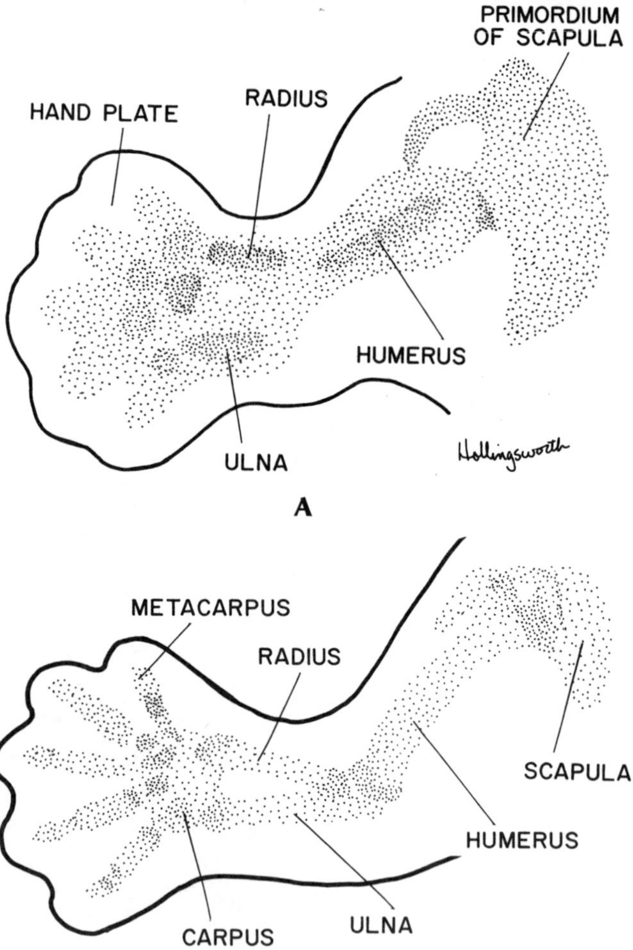

Fig. 57 Upper limb development: *A,* 4-week-old embryo; *B,* 6-week-old embryo.

borders which subsequently become the radial or thumb side and the ulnar or little finger side, respectively. A capillary plexus develops within the mesenchyme of each limb bud. This plexus joins intersegmental arteries which differentiate in the base of each limb bud; these capillaries eventually drain into a marginal vein. Ectoderm (apical) overlies the marginal vein, and it appears that both are essential to the mesoderm, which establishes the segments of the limb in a proximodistal sequence. Limb musculature, which differentiates in situ, is histologically identified as mesenchymal condensations near the base of each limb bud and around the ends of spinal nerves.

By the end of the fourth week, segmentation divides the limb bud into a shoulder girdle and an arm, forearm, and hand plate (Fig. 57A). The latter divides into finger rays in the fifth week. At six weeks (Fig. 57B), prior to bone formation, mesenchymal condensations, (scleroblastema) appear in a proximodistal sequence within the central axis of the limb bud.

It appears that for normal limb growth to occur, both limb mesenchyme and apical ectoderm must be present; moreover, the persistence of the latter is dependent upon an inductive factor that comes from the mesenchyme and is distributed asymmetrically. Thus, abnormal development may result from interference with this inductive mechanism, either by loss of response of the apical ectoderm or by failure of the inductor to reach and react with this ectoderm. Phocomelia may result when the mesenchymal cells, which give rise to the proximal limb bud segments, are damaged. Thalidomide can damage these cells and is effective only for a limited period of development (28 to 42 days ovulation age). Provided thalidomide is no longer active, normal distal limb segments may arise. That the upper limb buds develop slightly earlier than the lower may account for their greater frequency of damage.

Classification

How are limb malformations classified? To insure that malformations of limbs are recorded precisely and that registries and incidences of occurrence are accurate, nomenclature and classification must be standardized. Much of Saint-Hilaire's classification is still used (Fig. 58):

Limb defects as classified by Saint-Hilaire

1. Ectromeliens	phocomèle hémimèle ectromèle

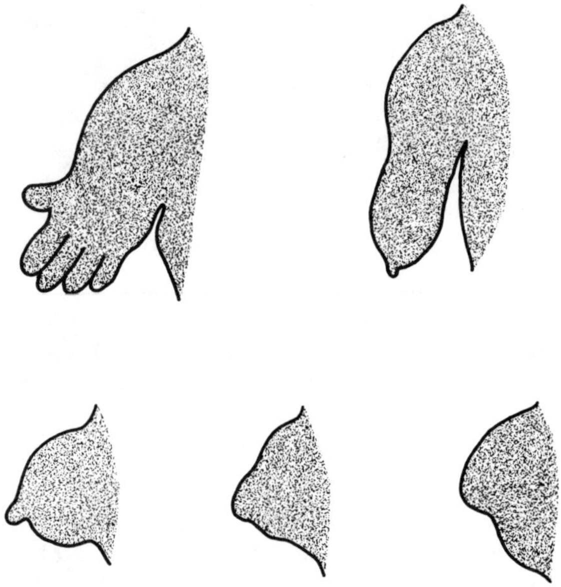

Fig. 58 Limb defects classified by Saint-Hilarie.

2. Symeliens — symèle, uromèle, sirénomèle

Frantz and O'Rahilly[8] in 1961 proposed a classification of congenital skeletal limb deficiencies (Table 5).

Table 5. Classification of congenital skeletal limb deficiences* ‡

Terminal (*T*)	
TRANSVERSE (–)	LONGITUDINAL (/)
1. Amelia (absence of limb)	1. Complete paraxial hemimelia (complete absence of one of the forearm or leg elements and of the corresponding portion of the hand or foot)—R, U, TI, or FI†
2. Hemimelia (absence of forearm and hand or leg and foot)	2. Incomplete paraxial hemimelia (similar to above, but part of defective element is present)—r, u, ti, or fi†
3. Partial hemimelia (part of forearm or leg is present)	3. Partial adactylia (absence of one to four digits and their metacarpals or metatarsals) : 1, 2, 3, 4, or 5
4. Acheiria or apodia (absence of hand or foot)	4. Partial aphalangia (absence of one or more phalanges from one to four digits) : 1, 2, 3, 4, or 5
5. Complete adactylia (absence of all five digits and their metacarpals or metatarsals)	
6. Complete aphalangia (absence of one or more phalanges from all five digits)	

Intercalary (I)

TRANSVERSE (—)	LONGITUDINAL (/)
1. Complete phocomelia (hand or foot attached directly to trunk)	1. Complete paraxial hemimelia (similar to corresponding terminal defect, but hand or foot is more or less complete)—R, U, TI, or FI†
2. Proximal phocomelia (hand and forearm, or foot and leg, attached directly to trunk)	2. Incomplete paraxial hemimelia (similar to corresponding terminal defect, but hand or foot is more or less complete)—r, u, ti, or fi†
3. Distal phocomelia (hand or foot attached directly to arm or thigh)	3. Partial adactylia (absence of all or part of a metacarpal or metatarsal) : 1 or 5
	4. Partial aphalangia (absence of proximal or middle phalanx or both from one or more digits): 1, 2, 3, 4, or 5

* *List of symbols:*

— transverse	I .... intercalary
/ longitudinal	R or r .... radial
: 1, 2, 3, 4, or 5 denote digital ray involved	T .... terminal
FI or fi .... fibular	TI or ti .... tibial
	U or u .... ulnar

A line below a numeral denotes upper-limb involvement; or example, T-2 represents terminal transverse hemimelia of the upper limb.

A line above a numeral denotes lower-limb involvement; for example, I-1 represents intercalary transverse complete phocomelia of the lower limb.

† In capital letters when the paraxial hemimelia is complete, in small letters when the defect is incomplete.

‡ From Frantz, C. H., and O'Rahilly, R.,[8] with permission.

In 1966 Burtch[4] revised this classification, and in 1969 Henkel and Willert[14] presented a classification of dysmelia with a pattern of limb malformation.

Associated Malformations

What defects most often accompany phocomelia? Obstructive lesions of the alimentary tract and cardiac and urinary tract malformations are frequently present.

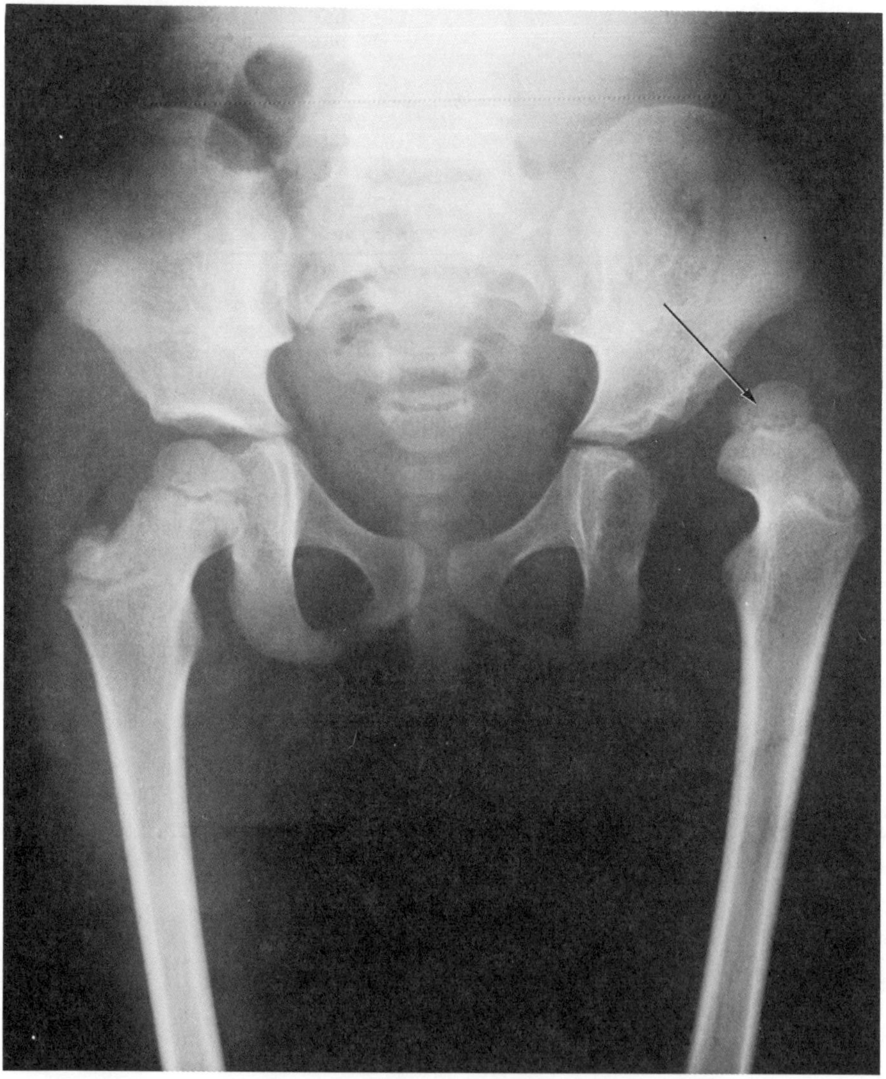

Congenital dysplasia with dislocation of the left hip.

31

Congenital Dysplasia of the Hip

Initial examination of a breech-born female infant, the product of an uncomplicated pregnancy of a gravida I, para I mother, revealed limitation of passive abduction (to 50°) of the left hip in the flexed position. The left side of the infant presented a positive Ortolani sign[24] and radiographs showed a sloping acetabulum, but were not helpful diagnostically due to absence of ossification centers. The family history revealed congenital dislocation of the hip, with four cases among close relatives.

Unequal thigh creases, unequal leg lengths, and external rotation of one leg are considered, at this age, unreliable signs of congenital hip dysplasia.

DIAGNOSIS

Congenital dysplasia with dislocation of the left hip (CDH).

THERAPY AND FURTHER COURSE

Prophylactic treatment was initiated, aimed at passive stretching of the adduction contracture and maintenance of the abducted and flexed position of the hips using a Frejka[9] pillow splint. The splint was worn continuously except during diaper changes, at which time gentle manual stretching of the hip in the direction of abduction was employed, until full abduction was achieved when the infant was 4 weeks old. Acetabular development was considered normal clinically and radiographically at 12 months and the splint was removed. At 4 years, the child's gait was normal.

DISCUSSION

Congenital dislocation of the hip occurs in 1/1,500 births[31] and has been recognized since the time of Hippocrates.[36] Pravaz (1837)[28] was the first to achieve successful reduction of CDH by a gradual method, and Lorenz (1895)[18] introduced separate methods of reduction and retention. Ortolani (1937)[24] showed that instability of the child's hip joint could be detected by clinical examination, and Frejka (1941)[9] introduced a splint that aided in reduction of CDH.

Embryology

Discuss development of the hip joint. All components of the hip joint differentiate in situ from one mesodermal mass (blastema). At 6 weeks of gestation (Fig. 59A) within this mesodermal mass, a zone (interzone) forms between the two eventual components of the hip joint, i.e., the pelvic bone and the head of the femur. Eventually, the interzone comes to surround the mesodermal mass in the area of the potential joint cavity.

By the seventh week (Fig. 59B) precartilaginous suggestions of the pelvic bone and head of the femur are found within the mesodermal mass.

In the eighth week (Fig. 59C) cartilaginous precursors of the entire lower limb and pelvic girdle are present. Concomitantly, the interzone divides into two areas: (1) the outer—which is continuous with the perichondrium of the bones comprising the hip joint; and (2) the inner—which forms the synovial membrane.

By the ninth week ossification centers appear within the ilium, and spaces within the synovial membrane appear and coalesce, forming the joint cavity. Dislocation cannot occur prior to formation of the hip joint cavity.

Etiology

What is the etiology of this congenital defect? Several factors have been suggested as the cause of CDH, e.g., poor development of the acetabulum, generalized joint laxity, intrauterine position, delayed development of otherwise normal structures, lack of tone of the joint capsule, and the newborn infant's reaction to maternal hormones that cause relaxation of this capsule.

Methods of Treatment

How would treatment of CDH differ in a newborn infant, a 1-year-

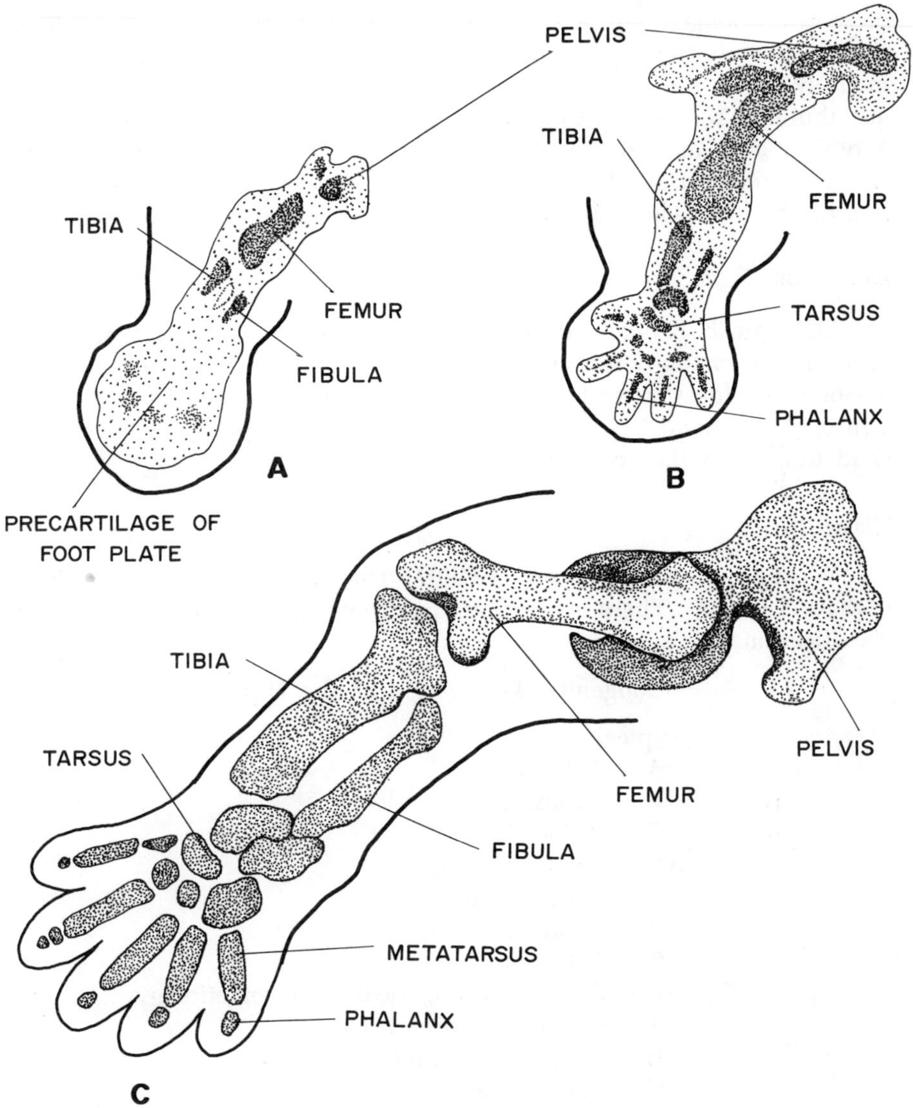

Fig. 59 Lower limb development: *A*, 6-week-old embryo; *B*, 7-week-old embryo; C, 8-week-old embryo.

old, and a 5-year-old child? Treatment in the newborn infant is prophylactic, aimed at passive stretching of the adduction contracture and maintenance of the abducted and flexed position of the hips by a pillow splint worn continuously for several months. In the 1-year-old child, the adduction contracture is more marked; there is apparent leg shortening, a protruding hip, and asymmetrical skin folds. The taut adductor muscles are lengthened and a closed reduction of the hip is maintained in a stable position by a plaster hip spica. The spica is changed bimonthly until radiographs reveal a satisfactory development of the acetabulum (six months to one year). At 5 years of age the child has a typical limp and marked bony changes of the hip. Treatment consists of a long period of traction, frequent tenotomy, and an open reduction of the hip. The major problem is maintenance of the reduction.

Definitions

Distinguish between dysplasia, subluxation, and dislocation. Dysplasia is a prenatal defect in development of the hip joint and is a precursor to subluxation and dislocation. Subluxation is an abnormal position of the femoral head in the acetabulum. In dislocation the femoral head lies out of the acetabulum.

Classification

What are the various types of CDH? The basic classification of CDH is that of Hass.[12] There are various modifications and additions to this classification.

Congenital Dislocation of the Hip

1. Typical
   - A. Dysplasia
   - B. Subluxation
   - C. Dislocation
2. Atypical (tetralogic)
   - A. Arthrogrypotic
   - B. Arthrocholarotic
   - C. Defective
3. Isolated cases of specifically known etiology
   - A. Traumatic (obstetrical)
   - B. Congenital syphilis
   - C. Little's disease
   - D. Roentgen injury

Associated Malformations

Congenital dislocation of the hip frequently occurs by itself or in association with other skeletal defects.

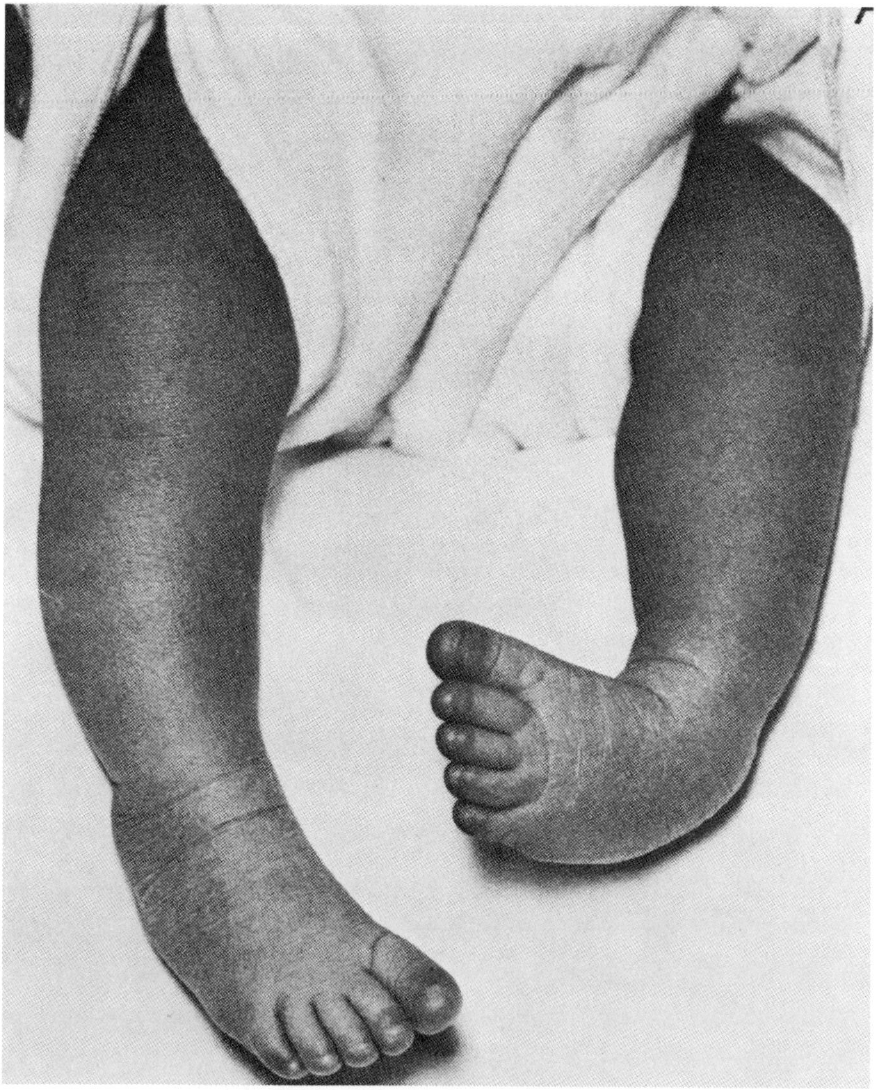

Left talipes equinovarus.

32

Clubfoot (Talipes)

A male infant was first seen in the newborn nursery for a left congenital clubfoot deformity with no additional malformations. Treatment was begun immediately. Weekly application of plaster casts to the left foot for gentle correction of the deformities (forefoot adduction, inversion, and equinus) was continued for nine weeks. The feet were then strapped to a Denis Browne boot splint (Fig. 60) for 12 weeks, with the affected foot progressively turned into valgus. The boot splint was then discontinued for longer periods until the child began walking, when it was employed only at night for one year. When the child was 2 years of age, straight-last boots were used for day wear.

DIAGNOSIS

Left clubfoot of the type talipes equinovarus. (Clubfoot is a congenital deformity of the foot, which is twisted out of shape or position).

TREATMENT AND FURTHER COURSE

When the child was 2 years old, there was difficulty putting the left foot into a shoe, and recurrence of the deformity was apparent. An anteroposterior roentgenogram indicated a loss of the normal angle between the talus and calcaneus (Fig. 61A, B); a lateral roentgenogram showed no dorsiflexion of the calcaneus. A tendo-achilles lengthening corrected the equinus deformity and the foot was positioned in maximum dorsiflexion. When the boy was 7 years of age, lateral and anteroposterior roentgenograms showed complete correction of the deformity.

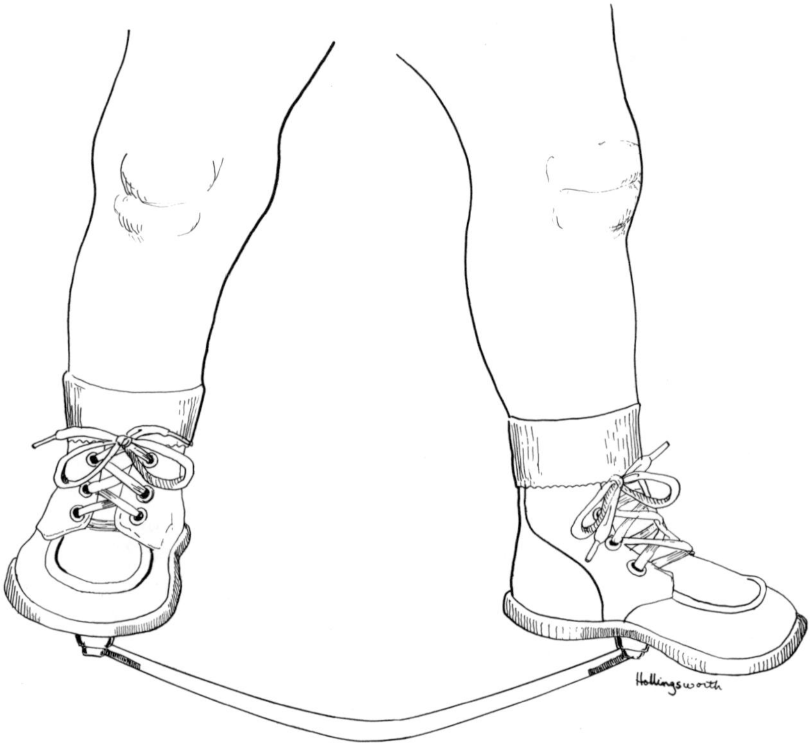

Fig. 60 Denis Browne boot splint.

Clinically, the feet were symmetrical. The boy wore regular shoes; the left foot was plantigrade and pliable, and dorsiflexion and plantar flexion of both feet were equal.

DISCUSSION

Hippocrates[20] is credited with the first description of clubfoot, a deformity that occurs in 1/1,000 births.[31] Hippocrates believed that the deformity resulted from mechanical causes, and that treatment should begin early in life. Other than in medical literature, clubfoot appears in novels and pictorial arts from the sixteenth to the nineteenth century.[30] Various aspects of clubfoot are difficult to understand and more difficult to solve, e.g., therapy and treatment, x-ray analysis, maintenance of correction, and etiology.[3, 7, 15, 17, 37]

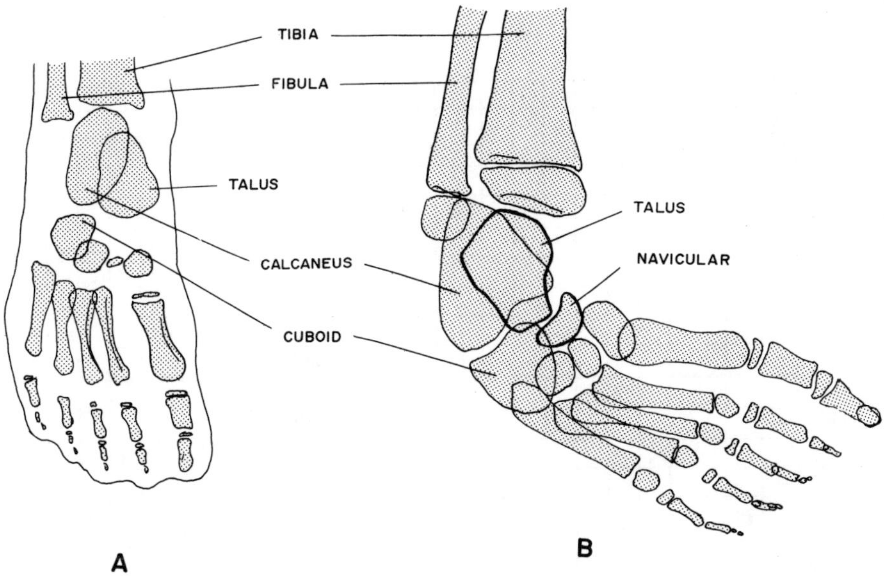

Fig. 61 Relationship of tarsal to metatarsal bones: *A,* Normal foot; *B,* Clubfoot.

Embryology

Discuss the development of the foot. In the fifth week of development, a paddle-shaped hind limb bud appears at the level of L_2 to S_3.

By the seventh week this bud begins to segment, first proximally and then distally into the hip, thigh, leg, and foot rays. Skeletal development proceeds in the same direction, with the bones of the foot forming about the connective tissue surrounding the plantar artery.

The digits of the foot are clearly defined by the eighth week, the lower limb assumes the fetal position, and the primary centers of ossification appear in the larger bones of the hind limb.

Ossification centers (Table 6) appear in the calcaneus and talus in the fifth and seventh fetal months, respectively. Only these two tarsal bones show ossification at the time of birth; moreover, it is the position of these two bones in relation to the metatarsals that determines the degree of deformity in clubfoot. Some metatarsal and phalangeal bones show ossification centers during fetal life; other bones of the foot possess ossification from the first to the fifth year postnatally.

Table 6. Appearance of ossification centers in the foot

Bones			Fetal Months									Years									
			1	2	3	4	5	6	7	8	9	1	2	3	4	5	6	7	8	9	10
TARSAL	Calcaneus	Body					X														
		Epiphyses																	X	X	X
	Talus								X												
	Cuboid										X										
	Navicular														X	X					
	Cuneiforms																				
	medial												X	X							
	intermediate													X	X						
	lateral											X									
METATARSAL	Body				X																
	Epiphyses													X	X						
PHALANGES	Proximal	Body					X					X	X	X							
		Epiphyses											X	X	X	X					
	Middle	Body								X											
		Epiphyses											X	X	X	X					
	Distal	Body		X																	
		Epiphyses											X	X	X	X					

Causes

What factors may contribute to the formation of clubfoot? Mechanical compression of the fetus in abnormal postures during the later months of pregnancy, nervous system malformations, spinal deformities, arrested development, and hydraulic pressure and congenital dislocation of the talocalcaneonavicular joint are theories used to explain congenital clubfoot.

Classification

What terms describe the variety of congenital clubfoot deformities? The following classification (Fig. 62) includes the majority of congenital clubfoot deformities:[5]

1. Talipes calcaneovalgus—heel is turned outward from the midline of the body and the anterior part of the foot is elevated.
2. Talipes calcaneovarus—heel is turned toward the midline of the body and the anterior part of the foot is elevated.
3. Talipes calcaneus—foot is dorsiflexed.
4. Talipes cavovalgus—longitudinal arch of the foot is dorsiflexed and heel is turned outward from the midline.
5. Talipes cavus—longitudinal arch of the foot is abnormally high.
6. Talipes equinovalgus—heel is elevated and turned outward from the midline of the body.
7. Talipes equinivarus—heel is turned inward from midline of the leg and the foot is plantar flexed (typical clubfoot).
8. Talipes equinus—foot is plantar flexed.
9. Talipes planovalgus—heel is turned outward from the midline of the leg and outer border of the anterior part of the foot is higher than the inner border.
10. Talipes valgus—heel is turned outward from the midline of the leg.
11. Talipes varus—heel is turned inward from the midline of the leg.

Associated Malformations

Spinal defects, neurologic impairment, and abnormal muscle insertions accompany congenital clubfoot.

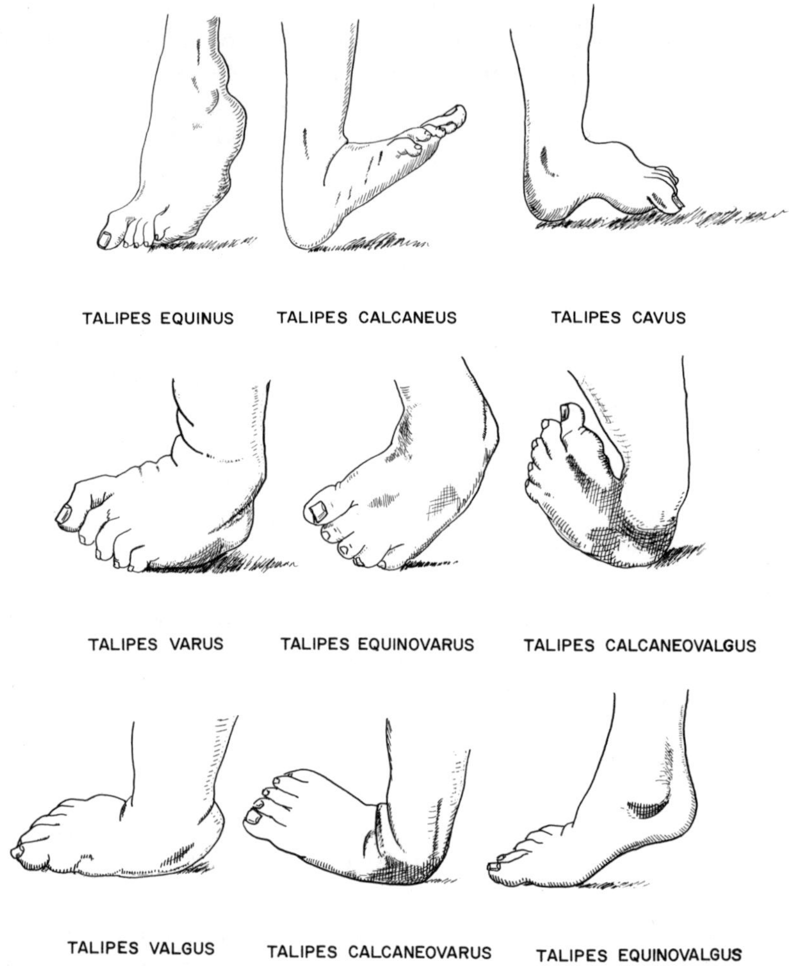

Fig. 62 Classification of congential clubfoot deformities.

REFERENCES

1. von Ammon, F. A.: Die angebornen chirurgischen Krankheiten des Menschen in Abbildungen dargestellt und durch erläuternden Test erklärt. F. A. Herling, Berlin, 1842.
2. Bartholin, T.: De naturae abundantia et defectu. Acta Med. Etophil. Hafn. 1673, 1675.

3. Browne, D.: The pathology and classification of talipes. Aust. New Zeal. J. Surg. 29:85, 1959.
4. Burtch, R. L.: Nomenclature for congenital skeletal limb deficiencies; a revision of the Frantz and O'Rahilly classification. Artif. Limbs 10:24, 1966.
5. Dorland's Illustrated Medical Dictionary, ed. 24. W. B. Saunders Company, Philadelphia, 1969.
6. Eagle, J. F., Jr., and Barrett, G. S.: Congenital deficiency of abdominal musculature with associated genitourinary abnormalities: Syndrome; report of 9 cases. Pediatrics 6:721, 1950.
7. Fait, M.: Suitable footwear for after treatment of pes equinovarus and other foot deformities. Acta Chir. Orthop. Traumat. Cech. 31:276, 1964.
8. Frantz, C. H., and O'Rahilly, R.: Congenital skeletal limb deficiencies. J. Bone Joint Surg. [Am.] 43:1204, 1961.
9. Frejka, B.: Präventation der angeborenen Hüftgelenksluxation durch das Abduktionspolster. Wien. Med. Wchr. 91:523, 1941.
10. Fröhlich, F.: Der Mangel der Muskeln, insbesondere der Seitenbauchmuskeln. C. A. Zurn, Würzburg, 1839.
11. Hamilton, S. G. I.: Congenital phocomelia in monozygotic twins. Brit. Med. J. 4:524, 1967.
12. Haas, J.: Congenital Dislocation of the Hip. Charles C Thomas, Springfield, Ill., 1951.
13. Henderson, B.: Congenital absence of abdominal muscles. Glasgow Med. J. 33:63, 1890.
14. Henkel, L., and Willert, H. G.: Dysmelia. A classification and a pattern of malformation in a group of congenital defects of the limbs. J. Bone Joint Surg. [Br.] 51:399, 1969.
15. Heywood, A. W. B.: The mechanics of the hindfoot in club foot as demonstrated radiographically. J. Bone Joint Surg. [Br.] 46:102, 1964.
16. Lenz, W.: Kindliche Missbildungen nach Medikament-Einnahme Während der Gravidität. Deutsch. Med. Wschr. 86:2555, 1961.
17. Lindermann, K., and Marquardt, E.: Historical review of therapy of congenital clubfoot. Z. Orthrop. 86:336, 1955.
18. Lorenz, A.: Die blutige Reposition der angeborenen Hüftverrenkung. Volkmanns Sommlung N.F. 117:169, 1895, Breitkopf and Hartel, Leipzig.
19. McBride, W. G.: Thalidomide and congenital abnormalities. Lancet 2:1358, 1961.

20. McCauley, J. C., Jr.: Clubfoot. History of the development and the concepts of pathogenesis and treatment. Clin. Orthop. 44:51, 1966.

21. Morgagni, J. B.: The Seat and Causes of Diseases Investigated by Anatomy: in Five Books, Containing a Great Variety of Dissections, with remarks translated by B. Alexander, Vol. 1. Millar, A., and Cadell, T., Johnson and Payne, London, 1769.

22. Mustard, W. T., Ravitch, M. M., Snyder, W. H., Jr., Welch, K. J., and Benson, C. D. (eds.): Pediatric Surgery. Year Book Medical Publishers, Inc., Chicago, 1969.

23. Nunn, I. N., and Stephens, F. D.: The triad syndrome: A composite anomaly of the abdominal wall, urinary system and testes. J. Urol. 86:782, 1961.

24. Ortolani, M.: Un segno poco noto e sua importanza per la diagnosi precoce di prelussazione congenita dell'anca. Pediatria (Napoli) 45:129, 1937.

25. Osler, W.: Congenital absence of abdominal muscles with distended and hypertrophied urinary bladder. Johns Hopkins Med. J. 12:331, 1901.

26. Parker, R. W.: Absence of abdominal muscles in an infant. Lancet 1:1252, 1895.

27. Pliny: Natural History Book VII, Cap. 2.

28. Pravaz, CH.A.: Du traitement de la luxation congenitale. Bull. Acad. Nat. Med. (Paris), 2:579, 1837.

29. von Recklinghausen, F.: Untersuchungen über die Spina Bifida. Virchow. Arch. Path. Anat. 105:243, 373, 1886.

30. Ruhräh, J.: Pediatrics in art, clubfoot. Am. J. Dis. Child. 46:1397, 1933.

31. Rubin, A.: Handbook of Congenital Malformations. W. B. Saunders Company, Philadelphia, 1967.

32. Saint-Hilaire, I. G.: Histoire generale et particuliere des anomalies de l'organisation chez l'homme et les animaux; ouvrage comprenant des reserches sur les caractères, la classification, l'influence physiologique et pathologique, les rapports généraux, les lois et les causes des monstruosités, des varietés et vices de conformation, ou traite de teratologie. J. B. Bailliere, Paris, 1832.

33. Silverman, F. M., and Huang, N.: Congenital absence of abdominal muscles associated with malformation of genitourinary and alimentary tracts; Report of cases and review of literature. Am. J. Dis. Child. 80:91, 1950.

34. Smithells, R. W.: Drugs and human malformations. In Woollam,

D. H. M. (ed.): Advances in Teratology. Logos Press, London, 1966.

35. Stumme, E. G.: Ueber die symmetrischen kongenitalen Bauchmuskeldefekte und über die Kombination derselben mit anderen Bildungsanomalien des Rumpfes. Mitt. Grenzgeb. Med. Chir. Jena 11:548, 1903.
36. Tuell, J. I.: The congenitally malformed. VI. Congenital hip disease. Northwest Med. 65:225, 1966.
37. Turco, V. J.: Surgical correction of the resistant club foot. One-stage posteromedial release with internal fixation: A preliminary report. J. Bone Joint Surg. [Am.] 53-477, 1971.
38. Walker, A. E.: A History of Neurological Surgery. Williams and Wilkins Company, Baltimore, 1951.
39. Williams, D. I., and Burkholder, G. V.: The prune belly syndrome. J. Urol. 98:244, 1967.

Index